INFORMATIVE PSYCHOMETRIC FILTERS

ROBERT A. M. GREGSON

Previous Books by the Author

Psychometrics of Similarity
Time Series in Psychology
Nonlinear Psychophysical Dynamics
n-Dimensional Nonlinear Psychophysics
Cascades and Fields in Perceptual Psychophysics

INFORMATIVE PSYCHOMETRIC FILTERS

ROBERT A. M. GREGSON

ANU
THE AUSTRALIAN NATIONAL UNIVERSITY
E PRESS

ANU
E PRESS

Published by ANU E Press
The Australian National University
Canberra ACT 0200, Australia
Email: anuepress@anu.edu.au
Web: http://epress.anu.edu.au

National Library of Australia
Cataloguing-in-Publication entry

Gregson, R. A. M. (Robert Anthony Mills), 1928- .
Informative psychometric filters.

Bibliography.
Includes index.
ISBN 1 920942 65 3 (pbk).
ISBN 1 920942 66 1 (online).

1. Psychometrics - Case studies. 2. Human beings -
Classification. 3. Psychological tests. 4. Reliability.
I. Title.

155.28

Cover design by ANU E Press.
The image on the front cover, that resembles a scorpion, is the Julia set in the
complex plane of the Complex Cubic Polynomial used as a core equation in Nonlinear
Psychophysics; it is employed in its equation form in some chapters in this book,
and is discussed in more detail in the author's three previous books on nonlinear
psychophysics and in some articles in the journal Nonlinear Dynamics, Psychology
and Life Sciences

Preface

There has been a time gap in what I have published in monograph form, during which some very important and profound shifts have developed in the way we can represent and analyse the sorts of data that human organisms generate as they progress through their lives. In that gap I have been fortunate enough to make contact with and exchange ideas with some people who have chosen to explore nonlinear dynamics, and to see if the ideas that are either pure mathematics or applications in various disciplines other than psychology can be usefully explored as filters of behavioural time series.

Three strands in my previous published work have come together here; they are time series, psychophysics, and nonlinear nearly chaotic dynamics. Also, the width in scale of the data examples examined has broadened to include some psychophysiological and social processes. There has evolved a proliferation of indices, hopefully to identify what is happening in behavioural data, that can be regarded as partial filters of information with very varying efficiency.

The title of this effort was selected to emphasise that what is measured is often information or entropy, and that the focus is on problems and data that are ill-behaved as compared with what might be found in, say, physics or engineering or neurophysiology. The object is to model in a selective way, to bring out some features of an underlying process that make some sense, and to avoid misidentifying signal as noise or noise as signal.

Recent developments in psychophysiology (Friston, 2005) have employed networks of mixed forward, reverse and lateral processes, some of which are linear and some nonlinear. That form of construction brings us closer to neurophysiological cortical structures, and takes theory further than is pursued here, though there is in both approaches an explicit assumption of nonlinear mappings playing a central role in what is now called the inverse problem; that is to say, working back from input-output data to the identification of a most-probable generating process.

There is no general and necessary relationship between identifiabilty, predictability and controllability in processes that we seek to understand as they evolve through time. In the physical sciences sometimes the three are sufficiently linked that we can model, and from a good fitting model we can predict and control. But in many areas we may be able to control without more than very local prediction, or predict without controlling, because the process under study is simple and linear, and autonomous from environmental perturbations. That does not hold in the life sciences, particularly in psychology outside the psychophysical laboratory.

The unidentifiabilty or undecidabilty in identification, prediction and control can be expressed in information measures, and in turn, using symbolic dynamics, can be expressed theoretically in terms of trajectories of attractors on manifolds. It is the extension of our ideas from linear autocorrelation and regression models to nonlinear dynamics that has belatedly impacted on some areas of psychology explored here.

A related problem that is unresolved in the current literature, for example in various insightful studies in the journal *Neural Computation*, is that of defining complexity. The precise and identifiable differences between complexity and randomness have been a stumbling block for those who want to advance very general metrics for differentiating the entropy properties of some real data in time-series form. I have not added to that dispute here, but sought simply to illustrate what sorts of complicated, nonstationary and locally unpredictable behaviour are ubiquitous in some areas of psychology. The approach is more akin to exploratory data analysis than to an algebraic formalism, without wishing to disparage either.

The problems of distinguishing between the trajectories of deterministic processes and the sequential outputs of stochastic processes, and consequently the related problem of identifying the component dynamics of mixtures of the two types of evolution, has produced a very extensive literature of theory and methods. One method that frequently features is so-called box-counting or cell-mapping, where a closed trajectory is trapped in a series of small contiguous regions as a precursor to computing measures of the dynamics, particularly the fractal dimensionality (for an example, which has parallels in the analysis of cubic maps in nonlinear psychophysics, see Udwadia and Guttalu, 1989).

Serious difficulties are met in identifying underlying dynamical processes when real data series are relatively short and the stochastic part is treated as noise (Aguirre & Billings, 1995), it is not necessarily the case that treating noise as additive and linearly superimposed is generically valid (Bethet, Petrossian, Residori, Roman & Fauve, 2003). Though diverse methods are successfully in use in analysing the typical data of some discipines, as in engineering, there are still apparently irresolvable intractabilities in exploring the biological sciences (particularly including psychology), and a proliferation of tentative modifications and computational devices have thus been proposed in the current literature.

The theoretical literature is dominated by examples from physics, such as considerations of quantum chaos, which are not demonstrably relevant for our purposes here. Special models are also created in economics, but macroeconomics is theoretically far removed from most viable models in psychophysiology. Models of individual choice, and the microeconmics of investor decisions, may have some interest for cognitive science, but the latter appears to be more fashionably grounded, at present, in neural networks, though again the problem of simultaneous small sample sizes, nonlinearity, non-stationarity, and high noise have been recognised and addressed (Lawrence, Tsoi & Giles, 1996).

One other important social change in the way sciences exchange information has in the last decade almost overtaken the printed word. For any one reference that can be cited in hard copy, a score or more are immediately identifiable in internet sources such as Google, and the changes and extensions of ideas, and perhaps also their refutation, happens at a rate that bypasses the printed text even under revisions and new editions. For this reason, there are some important topics that are not covered here, tools such as Jensen-Shannon divergence are related to entropy and to metric information and could well be used to augment the treatment of nonlinear and non-stationary psychological data but so far have not been considered. We urge the reader to augment and criticise the present text by checking developments in the electronic sources, particulary focussing on work such as that by Fuglede and Topsøe (www.math.ku.dk/topsoe/) on Jensen-Shannon divergence, or Nicolis and coworkers (2005) on dynamical aspects of interaction networks, that have relevance and promise. Jumps between modes of dynamical evolution even within one time series essentially characterise psychological processes, and transient states such as chimera (Abrams and Strogatz, 2006) may yet be identified in psychophysiology.

I want to thank various people who have encouraged or provoked me to try this filtering approach, and to bring together my more recent work that is scattered over published and unpublished papers, conference presentations, invited book chapters, and even in book reviews. One very congenial aspect of the modern developments in applied nonlinear dynamics is the conspicuously international character of the activity. Professors Stephen Guastello and Fred Abraham in the USA, Hannes Eisler in Sweden, John Geake in England, Ana Garriga-Trillo in Spain, and Don Byrne and Rachel Heath in Australia, have all offered me constructive help or encouragement over the last decade.

School of Psychology
The Australian National University

Contents

Chapter 1

Introduction

> The reader should be warned against being seduced into think-
> ing that linearization tells the whole story.

(Infeld and Rowlands, 2000, p. 296)

This monograph is about building models of psychological or psychophysiological data that extend through time, are inherently unstable and, even from the perspective of the applied mathematician, are often intractable. Such instabilities have not gone unnoticed by statisticians and even such patterns as good and bad patches in the performance of sports teams, which are inexplicable to some of their followers, have been modelled. Attempts to address this problem in a diversity of disciplines are legion and it is only data of interest to the psychologist and the constructor of psychological measurements that are our focus. This does not mean that new methods will not emerge, even while this monograph is being written. Strange attractors and soliton metamorphoses have been added to the range of theoretical constructs available to the physicist and the sorts of data that we may meet in psychometrics are often appropriately treated as evidence of non-coherence, though that term is, as yet, rarely used outside physics (Infeld and Rowlands, 2000). In one sense, psychological data can be even worse because they jump about in their dynamics or are induced to do so by the action of the environment providing stimuli. One might think of any attempt at modelling a real life process extending through time as a compromise between plausibility of a representation of substantive data and mathematical tractability. If that were so, it could be a zero sum game in which one is achieved at the expense of the other. But it is not. There can be and frequently is loss on both sides of the compromise.

Historically, we could go back to 1812, when one of the first serious mathematical models of sensory or cognitive processes was created by Herbart (translated by Gregson, 1998). He had the profound insight that

> The calculation of the rises and falls of imagery in consciousness this most general of all psychological phenomena, of which all others are all only modifications would only require a quite simple algebraic representation, if the imagery could be said to be directly proportional in all its strength, if not it has its origins in the perception of time itself, and would show against already existing contrasts.

But that does little more than remind us that the subsequent evolution of the discipline was a history of false starts and neglected solutions. There are two sorts of mathematical borrowing that are found in the history of quantitative psychology: borrowing of a wide area of mathematically expressed theory already in use in physics, biology or engineering, such as stochastic differential equations, or catastrophe theory; and borrowing of specific equations that were originally advanced as models in some substantive area that has no immediate intuitive parallels with psychology. The first sort of borrowing, if it can be called borrowing, is involved here and has been one of increasing interest recently as symbolic dynamics (Jiménez-Montaño, Feistel and Diez-Martinez, 2004). The second sort goes back a long way; for example, it seems not to be generally known that the use of a linear model with an added Gaussian random residual component, introduced by Fechner in the 1850s, goes back, via Weber, to Gauss and the resurveying of the streets of Hannover after the Napoleonic wars and to Gauss's monograph on least squares of 1809 (Bühler, 1986).

It seems to be relatively rare for mathematical psychology to produce its own equations that are not copies of something in physics or biology, such as borrowing the logistic equation created for population dynamics, but that may be an unfair criticism. If hard work has already been done, one can build on it. The complex cubic Gamma recursion of nonlinear psychophysics (Gregson, 1988, 1992), will be quoted later. It resembles some other mathematical models but is a bit different; a model can be one of a family and, at the same, uniquely applicable to some substantive area of science.

The title of this monograph was chosen to include the word *informative*, which has a special meaning that will characterise the approach taken: informative implies information, and information implies information theory and its extensions, associated with the names Kullback, Leibler and Akaike (de Leeuw, 1992). The use of information theory in psychology had a brief popularity, due in part to the efforts of Attneave (1959), preceding methods, in the later 20th century, that have had some partial impact on psychometrics but rather more on engineering. Recently, information measures as a basis of choice between alternative models has been strongly advocated in biology (Burnham and Anderson, 2002). This is an approach with which we have strong sympathy, but one that sets us at odds with some preferred traditions in psychophysics and in applied statistics. Much of both classical and modern psychophysics is written outside time. It is not an area of dynamics, let alone nonlinear dynamics, but one of steady-state stimulus-response mapping (Falmagne, 2002). There have been important attempts to extend this dynamically: Helson's (1964) work on adaptation level theory was one, and Vickers (1979) on accumulator theory was another. Trying to build time series analysis into the total picture that was mostly linear theory created this author's 1983 work *Time Series in Psychology*. But what has happened since creates a need for a fundamental rethink if nonlinear systems are to play a important role.

Ignoring sequential effects in stimulus-response mappings by only examining behaviour under asymptotic steady-state conditions is useful if some sorts of individual differences in responses to the environment are important; the logic of a simple intelligence or memory test does not ask how a respondent got, over some years, to be what he or she now is, but what he or she can now do when faced with tasks of various difficulties. One cannot investigate absolutely everything at once and deliberately choosing not to ask some questions is, perhaps paradoxically, one of the bases of scientific method. But asking too many questions at once, that is, collecting a host of data on all potentially relevant variables, can create studies in which nothing is identifiable or decidable. In statistics, this is called the problem of overdetermination or lack of degrees of freedom. The counter argument, particularly in nonlinear systems, is that, if variables are taken one by one and not in clusters as they occur in the real world, then the nature of their interaction is obscured and causality is obliterated.

Figure 1.1

$$\textbf{EVENT E} \quad \xrightarrow{f(E)} \quad Receptors\ R \quad \xrightarrow{f(E,R)} \quad \textbf{NEURAL}$$

$$\textbf{ACTIVITY N}$$

$$\Big\downarrow \phi$$

$$\downarrow \qquad\qquad\qquad\qquad\qquad\qquad\qquad \downarrow$$

$$Sampling\ g(E) \qquad\qquad \longrightarrow \qquad Sensory\ S1$$

$$\Big\downarrow \phi \qquad\qquad \overset{\psi\phi O}{\nearrow} \qquad\qquad \Big\downarrow \psi\phi(N,S1)$$

$$\textbf{INSTRUMENT} \quad \xrightarrow{F(g(E))} \quad Sensory\ S2 \quad \xrightarrow{\phi\psi(g(E))} \quad \textbf{OBSERVER}$$

$$\textbf{--ATION}$$

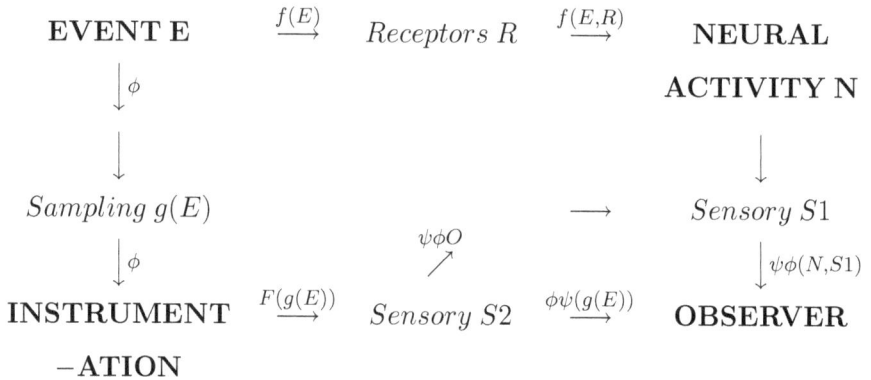

There are a number of locations in Figure 1.1 where a psychophysical mapping may be created. Indeed, in the 1860s, Fechner distinguised what he called psychophysics from below and psychophysics from above. In the 1890s, Wundt wrote as though he was predominantly seeing himself as doing physiology. It is explicit in Fechner (1860), that he wished

to bring together evidence in one framework from all of physics, physiology and everyday life. There are two pathways in Figure 1.1 that are
of interest: from the Event, through S1, to the Observer; and through the
Intrumentation of the Event and S2 to the Observer. The path through S2
is, as drawn, a bit of a cheat, as it should really go from Instrumentation
back to Receptors R and then via S1. The path via S2 is drawn to emphasise that some Events are never directly available to our senses, such as
voltages or radio waves (except in some pathologies, apparently), but we
can and do read pointer readings and may consider the psychophysical
relations between pointer readings and our awareness of events in ways
that were never available in the mid-19th century. Yesterday, as I am writing this, it was hot and muggy, so I consulted the barometer hanging in
the hallway of my home, and read humidity and temperature dials. Yes,
I was right. It was information consistent with feeling hot and stuffy; it
wasn't an illusion due to me having a bad cold. But I did not have a sound
meter handy to check the loudness of my hi-fi system. That just sounded
agreeable, though the nearby thunderstorm sounded threatening.

But there is an immediate complication: a step labelled 'sampling' indicates that only parts of the Event get represented in the Instrumentation
and some of the properties encoded may never correspond to our sensory experiences. This is particularly true for detailed chemical analyses,
as compared with odours and tastes. Unprocessed signals can come as far
as S1 and/or S2, but get no further.

The psychophysical mappings, symbolised in Figure 1.1 as $\psi\phi$ terms,
and expressed in equations; traditionally they are usually called laws,
named after Weber, Fechner, or Plateau (but his was later reinstated by
Stevens). These equations are peculiar, for two reasons: they are usually incompletely expressed, as they have no stipulated boundary conditions but
are intended only to hold within a limited Event amplitude range, called
lower and upper thresholds; and they treat the Event as fixed in time, sufficiently encoded as a scalar variable. There is no provision for dynamics
in the environment, but some for delays in the sensory pathways. Fechner
regarded the method of average response to a fixed stimulus as one for getting a best estimate from repeated presentations of a fixed environmental

stimulus that created local second-order variability in the sensory system, though the dynamics of neither were then of interest, only the first and second moments. Much of contemporary psychophysics does just that.

If one wanted to rewrite Fechner's version of Weber's law with boundaries, then for Response = ψ, Stimulus = ϕ,

$$f(\Delta^1 \phi | p = .5) = a \cdot log(\phi), \qquad \phi_{min} < \phi < \phi^{max} \qquad [1.1]$$

might be acceptable. Note that the equation is strictly not psychophysical, in that psychological units of magnitude are replaced by a j.n.d. operation expressed in physical units. There are serious problems with the rigorous interpretation of [1.1] as a scale of measurement, that are discussed by Luce and Galanter (1951).

The alternative popularised by Stevens (1951) is

$$\psi = a \cdot \phi^b, \qquad \phi_{min} < \phi < \phi^{max} \qquad [1.2]$$

and this form does map from physical to psychological measures, but the choice of what actually to use as ψ is very arbitrary.

As both [1.1] and [1.2] are monotonic and, as written, have no additive constants, one can be transformed to the other. There are objections to [1.2] concerning what happens in the region near ϕ_{min}. Those objections go back to Wundt. In order for an observer to perform the tasks necessary to create data for either equation, some sort of comparison of successive events is needed, but the events are taken as being independent realisations of the same stationary process. To put it another way, they are both models of relative judgement or, as Luce and Galanter subsumed it, in the general category of discrimination tasks. What is also interesting here is that neither equation describes a process that is self-terminating in time, whereas in nonlinear dynamics that can be done without adding independent boundary conditions. What are treated as events are not themselves instantaneous, but the temporal extension of the events may not feature in [1.1] or [1.2], unless the event is itself a time interval whose duration is to be judged. The time to process in S1 or S2 may be much longer than the temporal duration of the event, as is involved in short-term memory, or the delays in the instrumentation and it is this difference in processing duration that commonly characterises the whole system. A consequence is that separate equations for response latencies, or reaction times as they were sometimes called, have to be added to the system. A special case that arises in audition is the perception of pitch; the signal is fluctuating in time and the eventual response is a single unfluctuating level, the Neural Network in the ear integrates the fluctuations in a time sample, so the event must have duration. We will explore these filters in later chapters.

The pathways in Figure 1.1 can be thought of as imperfect information transmitters in various senses: the instrumentation only detects some Event properties, both intentionally and unintentionally; the neural networks that include sensory channels to the brain have limitations in activity level capacities and rates of transmission; and the eventual conscious observer has fluctuations in attention and memory, both short and long term.

Filters and Filtering

There is a story, probably not apocryphal, that the great British chemist Joseph Priestley (1733-1804) cut a hole in the base of his door so that his cat could come and go through it[1]. This in itself is not exceptional, today people do it with a little hinged panel on the hole and it is called a cat-flap. But Priestley also cut, alongside the larger hole that was big enough for an adult cat but too small for a dog, a much smaller hole for a kitten. This is seen as eccentricity, but can also be the result of shrewd observation of cat behaviour: a mother cat will carry a very small kitten by holding the back of the kitten's neck in her jaws, older kittens who can see and run will run alongside their mother where she can keep an eye on them, not necessarily follow her from behind, out of her sight. Priestley was a careful observer

---

[1] This story is also attributed to Sir Isaac Newton, who is said to have cut not one hole for one kitten but four holes alongside each other, for the whole litter. The principle of the story is unaltered but, as recent scholarship portrays Priestley as a nicer person than Newton, let us give him some credit.

and a prolific theorist and could have argued that a cat with a kitten fleeing a predator would have the best chance of survival if they could run in parallel. The oddity to some is that the kitten would grow too big for the small hole and change its behaviour, if it were still around, so that it could follow or lead another cat through one large hole. What we are going to call a filter is exemplified by Priestley's cat hole. Its purpose is to let one thing through but not another, and has to be modified as circumstances change. Those changes may or may not be predictable and, if they are not, the filter may become ineffective.

This monograph is about time series in psychology, which means it is about records of sequences of events in life histories. At any arbitrary point at which an external observer stands, the information about what is happening in the time series can change or can remain the same. If the changes are partial, some features persist and others are lost, then, by definition, in the sense that we are going to use the word *filter* here, a filter is operating. Any filter is an operation that lies between two uninformative extremes: the equifinality of death on the one hand, and perfect transparency on the other. One extreme ensures that no information survives the moment of observation, the other tells us absolutely nothing new. But both extremes are perfect bases for a type of simple prediction; that makes the point that predictability is not the same thing as being able to be filtered, but filtering is a form of partial prediction.

A filter can preserve information, it can destroy information, it can create false information, if the action of the filter is encoded in some way in information measures. If a metric can be imposed on information in a process, then the relative contributions of preservation, destruction and creation may be quantified, and then risk assessment attendant on information loss or misrepresentation can be determined within limits. Priestley's cat hole did not function quite the same way as the filters considered here, but it shares some features in the decisions it imposed on and provided for the cat and her kitten.

There is an important distinction that can be made between a *filter* and a *sieve*. At the same time, a sieve is a particular form of filter. If there exists a basis for ordering component objects or signals in magnitude, usually but not necessarily on a single continuum, then those magnitudes are the basis on which a sieve separates the set of objects into two or more subsets and allows one to pass through in time. The idea of a sieve in a physical sense is thousands of years old, wheat is separated from chaff, gold is panned by swirling out less dense particles from sand or gravel. But when information in time series is being transmitted, it is usual to write of *band pass filtering*, a term that covers a diversity of possibilities. Then, events that lie in a narrow magnitude band are given a special status, either for acceptance or rejection. The bands are often defined in terms of frequency components, as is common practice in electroencephalogram analyses. What is not passed on may be defined as noise. If very slowly or very quickly fluctuating components in a mixture of components of a time series are given a priori some special meaning, then they are filtered in or out by using, respectively, high- or low-pass filters. The following diagram (Figure 1.2) is a simple heuristic to relate some parts of filtering.

There is another sense in which information is filtered: censoring. The name obviously derives from its older usage in distasteful politics or government, but has been borrowed by statisticians to indicate when a frequency distribution of data has been in some way truncated or trimmed, usually to exclude a long tail or outliers. This obviously implies that there is some prior notion of what the full frequency distribution would be like and, as a consequence of censorship, estimates of parameters are suspected or known to be biased.

The flow diagram in Figure 1.2 is skeletal, it includes most of the information flows that will concern us and where they are but, as it stands, would be useless in exploring any real situation, because it has no equations, variables, parameters, or gains at any point and no specified delays in the pathways. All these are necessary in order to create a simulation and compare that simulation against real world events.

The words IN and OUT indicate places where the system would be externally observable. *Externally* refers to any instrumentation, either behavioural or psychophysiological, that is accessible independently of the assumptions in the system representation. A symbol δ indicates variable delay in feedback (fb) that is controlled independently of the intrinsic de-

Figure 1.2

$$
\begin{array}{ccccc}
 & Model & & & \\
 & First & & & \\
\mathbf{IN} & Version & & Noise\longrightarrow & \mathbf{OUT} \\
\downarrow & \downarrow & \nearrow & & \\
Time\ Series\ Data \longrightarrow & \mathbf{FILTER} & \longrightarrow & Information\leftarrow & fb \\
\downarrow & \uparrow & & \downarrow & \uparrow \\
\downarrow \mathbf{OUT} & New\ Filter & \swarrow & Prediction & \uparrow \\
Stationary & \uparrow & & \downarrow \mathbf{OUT} & \uparrow \\
Subset \longrightarrow & Tests & \longleftarrow & Disparities & \uparrow \delta \\
 & \uparrow \mathbf{IN} & \swarrow & \downarrow \mathbf{OUT} & \uparrow \\
 & Iterated & & \downarrow & \uparrow \\
 & Model \longrightarrow & & Control\longrightarrow & fb \\
 & & & InRealTime & \\
\end{array}
$$

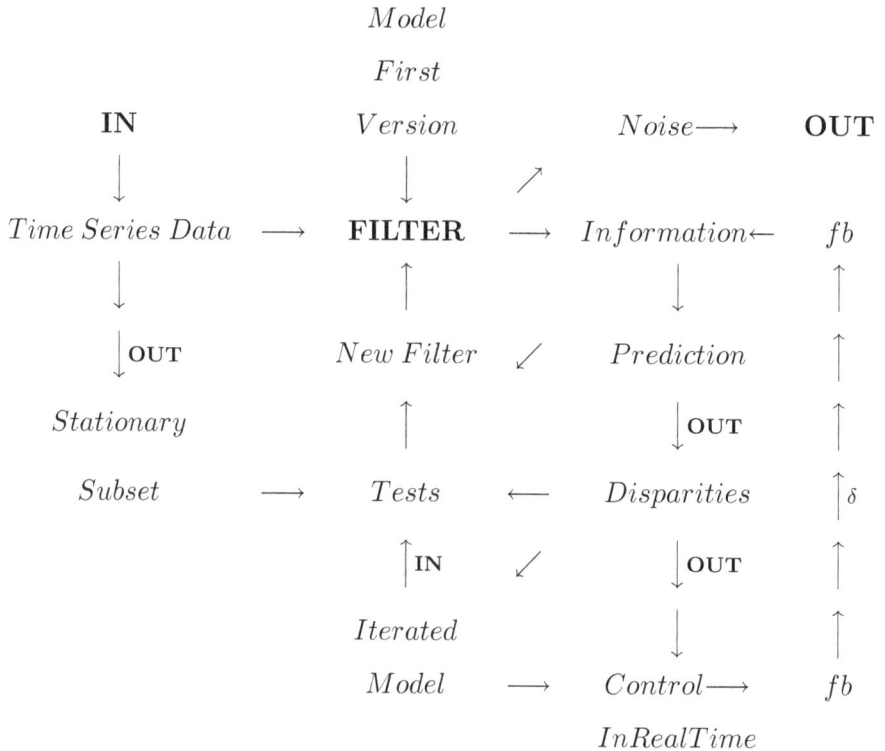

lays of the filtering operations themselves.

Many of the steps in the flow diagram are, to some extent, discretionary and involve decisions about when to stop an iterative revision of modelling and prediction. Such iterative processes that should converge, if the analysis is stable, but might not converge on a useful solution (hence the word "applicable" is implicit in the title of this monograph), are themselves the subject of computational filtering when they become written as computer programs. One discretionary step is the abstraction of a subset of data that, on inspection, appears to be stationary with respect to the lower-order variables of the model being used as the central filter. Lower-order here means the parts of a model that are treated as constant locally in time, so in turn implies taking short subsamples. As always, the problem of non-stationarity is ubiquitous in psychological data.

It is assumed in the flow diagram that the input data that is externally observable is complete in terms of the inbuilt assumptions of the initial model. However, many data sets are censored, and particularly may be *aliased*. This term refers to yet another sort of filtering and is most often used to denote a situation where high frequency components of a series encoded in the frequency domain are lost because the frequency of discrete encoding is less than twice the maximum frequency of the signal's components. But aliasing can also arise in a different way; for example, the statistics of the long series of major accidents resulting in deaths in English coal mines over the century beginning around 1850 does not include accidents where there were fewer than 10 men killed at once. Aliasing small accidents seriously biases both the distributions of accident magnitudes and the inter-accident intervals in time. This confounds our understanding of the relevant causalities. There is no difficulty in imagining similar problems in self-report records or recall by human subjects who keep diaries of their misfortunes, moods, or illnesses.

The filtering that yields both information and noise is, in experimental psychology, most commonly that created by the General Linear Model (GLM), of which ANOVA is a special case. The part that is filtered out as noise is assumed to be Gaussian, to have independent identically distributed (i.i.d.) realisations and be independent of the linear model components. In general, these are empirically false but mathematically tractable assumptions. The GLM is not usually applied iteratively in experimental psychology. The assumptions made in the GLM will not be used here, with a few exceptions.

A partitioning between signal and noise is traditional in linear models under stationarity, but it does not follow that analyses in the frequency domain, when cyclic patterns are observed or suspected, are necessarily most advantageous. An interesting example of using eigenvector decompositions of autocovariance matrices was described by Basilevsky and Hum (1977); it has various advantages if the time series variables have some metric properties and can be used for weak stationary short series. It also has the advantage that series need not be detrended prior to analysis and the partitioning of causal components can be easier to interpret. For these reasons, we will return to eigenvalue methods in some cases in later chapters, even though we will need to relax assumptions about metric variables and stationarity.

The input time series may be multivariate and the measures that comprise it may be real or complex if they are apparently numerical but, more generally, are symbols, encoding different types of behaviour such as are used by ethologists in studying animal behaviour patterns and sequences. Priestley's cat and her kittens need representation in a series of behaviour sequences, parallel in time and space, and assumptions of interdependence of component series can be erroneously filtered out. Non-numerical symbols are still information and can still be filtered, though the filtering may then not be possible in the frequency domain. If the initial model makes wrong assumptions about the properties of the input series, then the filter will not be optimal, as the filter is an extension of the model, built on the same assumptions about the nature of the input. A multicomponent time series may be misidentified and wrongly filtered if assumptions about the mutual independence of parallel input component channels are wrong.

Special cases, that arise in nonlinear dynamics, where the input series is treated as a trajectory in a dynamical system, may also be modelled as recursive steps within a neural network. In such cases, the variables can be complex (Hirose, 2003).

Given the suspected existence of a dynamic system in the real world, it is possible to proceed through a series of ordered stages, with success or failure at any stage. Any step may be partially implemented, and the degree of partial implementation can be sufficient to progress to the next stage, with a consequent increasing risk of failure.

It is also possible to achieve some later stage, such as Control, without, in fact, having any predictive capacity beyond the short-term.

[1] Observation

This refers to ordinary language observations of events and properties of events with minimal theoretical presuppositions. Let us call these the set Ω. One might call this stage extreme atheoretical empiricism.

[2] Description

It is assumed that some framework or vocabulary for grouping and ordering events is employed, which takes some account, not necessarily valid, of the relative substantive a priori significance (not necessarily statistical significance) and reliability of different data components, when they are potentially to be incorporated in a total description of the situation under examination. The descriptive attributes imposed by the observer are the set $a, b, c, d, \ldots$ and so Ω becomes $\Omega_{abcd\ldots}$.

[3] Measurement

This is defined as the first stage in which observations are replaced by numerical or symbolic encodings, as those numbers $x, y, \ldots$, with assumed metric properties of some type $\mu(x, y, ..)$, are then the working basis of the next stages. This is the step at which ordinary language meanings cease to feature if the encoding is consistent and unambiguous. This consistency refers to the mappings $\Omega_{abcd\ldots} \mapsto \mu(xy..)$. In general the mapping will be many-to-one, multiple instances of events classified as identical in terms of their attributes will map into one set of numerical variables. This implies a condensation of structure prior to formal modelling.

[4] Modelling

This is defined as the formal construction of mathematical or symbolic programming structures, which are comprised of parameters and variables and the relations between them. This definition is deliberately vague, as it covers any model that can be written as code, in digital or analogue mode, and would thus include any differential-difference equations and recursive maps. The purpose here is to focus on models that are intended to reflect the putative properties of system dynamics. If the set of relational operators in a model M is $\{\rho_M\}$, then

$$M \equiv_{def} \rho_M, \Omega_{abcd..}$$

[5] Identification

When, by some incomplete process of matching a subset Q of the properties of M is matched down to a residual error limit ϵ with a real data set not used in [4], the process is said to be Q, ϵ-identified. This idea parallels the statistician's notions of goodness-of-fit and the costs of misidentification.

[6] Prediction

A prediction is a statement at Q, ϵ level about a set of events [1] or [2] that have not been employed in the construction of M. They may be future events, or ones that are already on file but not employed in [4,5].

[7] Control

In order to write about control it is necessary to define first the notion of acceptable limits of variation within a Q, ϵ system. We assume that a representation $\hat{M}$ exists and that it can, over time, drift out of satisfying Q, ϵ conditions for M. To control effectively is to add a set of rules for the transient implementation of a feedback loop that was not originally part of M; either current variable values are changed by being overwritten, or parameters are reset. If the control is recursive, then the need for prediction is minimal. Otherwise, local predictability is required. Control involves the use of energy and thus may reduce the efficiency of a system which embodies M.

[8] Response

All of [1] through [4] may be incomplete and lead to pseudo-identification in [5]. Importantly, in [1], if we are to consider dynamics, then already observations are tagged as occurring in time and in temporal separation, coincident or sequential. By Response [8] is meant the capacity

of the observer to take protective or corrective action to change the outcomes of the action of a system that cannot be controlled. For example, the
weather can be predicted in the very short term, not be well predicted in
medium term, and be well predicted on average with seasonal correction
in the long term. The response is sometimes to carry an umbrella or to use
a sun-screen cream.

The change from [1] to [2] involves the use of some categorisation of
raw data, so a priori notions of similarity and difference and clustering become involved. When the partially described data are quantified, either by
direct measurement of properties in extension, or by probabilities of class
membership, then we move to [3]. Modelling [4] uses [3] and operations of
identity and relations. The errors arise at this stage either because of wrong
variables being included at [2,3] or wrong relationships being postulated
in [4]. Dynamic models involve relationships that evolve in time.

All psychological data are samples from the evolving life histories of
the observed individuals who are the subjects of observation and theorising. From the perspectives of time series analysis they generate data whose
numerical properties are ill-behaved or even ill-defined, and in terms of
linear models they are non-stationary unless some ruthless data smoothing or filtering is applied locally. The wider the windows in time through
which we observe the behaviour, the more insight we potentially obtain,
but the less tractable the data become. Predictability means extrapolation
into future time of dynamic sequences whose structures are incompletely
observed, and whose instabilities can be shown to be only second-order.

Time series fall into different types: event series, time-interval series,
and events-in-time series. Further subdivision of these types is possible,
according to whether or not the variables are continuous or categorical,
and whether the observed events are treated as seen through narrow windows opened upon an underlying continuous process (Krut'ko, 1969) or
are assumed to be discrete events exhaustively recorded. The statistical
methods used to identify sequential dependencies in the dynamics can be
quite different for the various types.

Any series which is representable in the time domain can also have an
equivalent representation in the frequency domain using Fourier analysis

and, computationally, the Fast Fourier Transform. That approach yields a representation of the relative energy in any frequency component of the process and is used widely in EEG analyses, which are typically event series of potential levels recorded at 3 msec intervals, when a physiological significance is given to energy in the 10-12 Hz spectral band.

Event Series

An event series is here defined as a series of events equally spaced in time. Event x_j arrives at time τ_j, but the temporal units in which τ is recorded play no part in the data analysis, though they play a part in the subsequent interpretation of the process. The first differences of the time process $\Delta^1 \tau_j$ are zero or taken as constant, or are assumed to be sufficiently near constant not to be important in their variation. For example, observations are made once a month on total suicides in a community, or once every ten seconds on attention shifts to a television program, or once every ten msecs to an EEG measure. The time series analysis is applied to the sequence,

$$x_1, ..., x_j, ..., x_n$$

which is taken to be unbroken, though missing values which are randomly distributed and do not exceed more than about 5% of n can be filled by local linear interpolation, matching iteratively the first two moments of the distribution of x. The widespread use of ARIMA modelling introduced by Box and Jenkins (1970) is almost always concerned with event series as defined here, but modifications to handle local discontinuities and interpolated episodes are useful in psychological applications (Gregson, 1983).

The results of analysis are expressed in autocorrelation spectra and then models of minimum order are fitted, within the general framework of (p, d, q) where p is the order of the autoregressive parts of the model, d is the order of differencing ($\Delta^d(x)$), and q is the order of the moving average parts of the model. We shall not be using much of that theory in this work because the time series encountered in psychology are often too short or too nonstationary in ARIMA terms to legitimise the methods beyond exploratory data analyses, so the reader is referred to standard sources such as Hannan (1967) or Kendall (1973).

Time-interval Series

When the occurrence of an event, but not its magnitude, is considered to be the meaningful basis of data, and the events do not arrive equally spaced in time but with inter-event-intervals (iei) $d\tau_j$, the frequency distribution of $d\tau$ is the focus of statistical analysis. The iei series is itself a time series and may be explored using all the methods available for an event series, the interpretation of results is, however, quite a different matter.

Events-in-Time Series

It should be immediately obvious that series in which both the intervals between events have a distribution in real time and the events themselves vary in magnitude occur in the real world. In fact, one would think that psychological data are preferably recorded and encoded in this form, though examples are, in fact, sparse. A series of judgements of sensory intensity may be made, each with a response delay (or reaction time) after stimulus presentation and the arrival of the stimuli may be random or locally unpredictable. For example, in a vigilance task, looking at a radar screen for the movement of objects in airspace, some signals may be seen quickly, some after a delay, and some missed completely. The objects generating the signals and the signals themselves can vary in size, proximity to the observer, and rate of movement.

In the notation just introduced for the two previous cases, each event is a pair $x_j, d\tau_j$ and this series may be represented in complex variables,

$$x_j, id\tau_j$$

The advantage of using complex variable models for the generation of this process is that the imaginary part may exhibit fast dynamics at the same time that the real part changes relatively slowly. Fast/slow dynamics appear to be widespread in psychophysiological contexts (Gregson, 2000), so the possibility of treating event-in-time series as complex and not as two simultaneous reals needs a priori consideration.

Markov and Hidden Markov models

If data are only encoded in an exhaustive mutually exclusive set of categories, which need not be ordered, then the event series and events-in-time series can be represented in another fashion. The variable x_j may be a label of a set of k ordered states, or it may be the result of a coarse scaling in which x is partitioned into k ordered categories which span the total range of x. The latter approach is particularly useful where the response data in a psychological experiment are not more than steps on a Likert scale.

The core of any k-state Markovian model is the transition probability matrix $\mathbf{T}_{k \times k}$, which is taken as having fixed elements in a stationary realisation. It thus follows that variation in estimated transition matrices over successive subseries are an empirical test of non-stationarity.

$$\mathbf{T}_{k \times k} = \begin{pmatrix} p_{11} & p_{12} & p_{13} & \cdots & p_{1k} \\ p_{21} & p_{22} & p_{23} & \cdots & p_{2k} \\ \vdots & \vdots & \vdots & \ddots & \vdots \\ p_{k1} & p_{k2} & p_{k3} & \cdots & p_{kk} \end{pmatrix}$$

The convention of interpretation is that the rows are the jth trial states $1, ..., k$ and the columns are the $(j+1)^{th}$ states so that the transition probabilities are the system's one-lag stochastic dependencies. If it is possible for transition to happen from any one state to any other over a finite series of trials, the process is ergodic, and the stationary k-state vector $\mathbf{V}_\infty$ is computable. That is,

$$\exists \mathbf{V}_\infty \quad such \ that \quad \mathbf{V}_\infty^{'} = \mathbf{V}_\infty \mathbf{T}$$

The extension of this model to what is called a hidden Markov process arises when the dwell time in any state is not one lag with an exit probability vector given by the kth row of $\mathbf{T}$, but is an exponential decay function, where the probability of staying in the state over more than one time unit decreases with a decay parameter specific to that state. That is, k more parameters are added to the model. This case has so far had negligible use in psychometrics, but can be usefully regarded as an example of fast/slow dynamics, where the fast dynamics are within a state and the slow dynamics are controlling the inter-state transitions. It becomes intractable as an estimation problem for short series with a large number of states or long series with a low number of states (Visser et al, 2000). As the focus of this study is on short series with a priori an unknown number of states, it is set aside. It is the failures of the simplest Markov model, where the probability distributions of lengths of runs in a state are not the power series $p_{jj}^s, j = 1, ..., k, s = 1, .., \infty, p < 1$ but are more persistent in some states, that are of particular interest in psychology.

Stimulus-response sequences

A special case, which involves multivariate time series and particular interpretations of the components of the vector $\mathbf{x} \in \mathbf{X}$ at any stage j, arises in psychological models and is an extension of the previous cases. Its picture has been called an influence-lines diagram (Gregson, 1983), which is useful for revealing the hidden ambiguity in many experiments between stimulus-dependent and response-dependent sequential processes. The diagram is a window of fixed length on the process, which may extend to an indeterminate degree outside the window both in the past and in the future. In the statistical sense, it is sometimes called a moving-boxcar window, and the process is stationary if the implicit scalar weights on all the influence lines within the window remain the same; that is to say, they are independent of the time counter j. We use $\mathbf{U}$ to mean an uncontingent process and $\mathbf{C}$ a contingent process. Here, j is the time sequence counter, which in an experiment is a trial number, s is a stimulus magnitude, r

a response magnitude, and er is an expected response. In fact, er can be any ancillary psychological variable, such as subjective confidence ratings or response latencies, or even a measure of surprise. This latter ancillary variable can run into future time, whereas, except in the dubious context of parapsychology, s and r do not. The vector $\mathbf{X}_j = (s_j, r_j, er_j)$.

In $\mathbf{U}$ the stimulus series is externally defined, and may be generated by a function $s_{j+1} = f(s_j)$, where f may be deterministic or stochastic; in the latter case we may write $s_j = f(s_{j-1}, s_{j-2}, ..., s_{j-m}, \epsilon_j)$, $\epsilon \sim N(0, \sigma)$. The lag parameter m may be unknown or may be defined by the experimenter.

$$
\mathbf{U} := \begin{pmatrix}
past & & past & & \mathbf{now} & & future & & future & \\
s_{j-2} & \rightarrow & s_{j-1} & \rightarrow & s_j & \rightarrow & s_{j+1} & \rightarrow & s_{j+2} & \rightarrow \\
\downarrow & \searrow & \downarrow & \searrow & \downarrow & \searrow & & & & \\
r_{j-2} & \rightarrow & r_{j-1} & \rightarrow & r_j & \rightarrow & \cdots & \cdots & \cdots & \cdots \\
\downarrow & \nearrow & \downarrow & \nearrow & \downarrow & \nearrow & \cdots & \nearrow & \cdots & \cdots \\
er_{j-2} & \rightarrow & er_{j-1} & \rightarrow & er_j & \rightarrow & er_{j+1} & \rightarrow & er_{j+2} & \rightarrow
\end{pmatrix}
$$

$\mathbf{U}$ may be regarded as the normal paradigm for a psychophysical experiment, though most usually only the set of vectors $\{s_j, r_j\}$ is recorded, and the mapping $r = \Phi(s)$ is of interest, outside of time. The terms er many be replaced or augmented by subjective confidence ratings, particularly if the task defining $s \rightarrow r$ is identification and not estimation.

There is another interpretation, if the task is to learn some associations between s and some outcomes e^*r which are provided by the experimenter (and not by the subject) after r, in some cases generating surprise (Dickinson, 2001). In this case the e^*r have the role of conditioning stimuli in classical conditioning. Rescorla and Durlach (1981) distinguished between within-event and between-event learning: the former needs only s_j, r_j pairings, whereas the latter requires the sequential linkages $j - k, ...j - 1, \mapsto j$ in the diagram.

An extended psychophysical function may be defined, where m and n are unknowns, as

$$
r_j = \Phi^*(\{s\}, \{r\}) = f(s_j, s_{j-1}, ..., s_{j-m}, r_{j-1}, r_{j-1}, ..., r_{j-n}, \epsilon_j)
$$

and we note that statistically, if it is linear, Φ^* is an ARMA process.

The contingent process **C** is one in which the stimulus series is independent of the environment and is internally-generated contingent on the previous response series.

$$
\mathbf{C} := \begin{pmatrix}
past & past & \mathbf{now} & future & future & \\
s_{j-2} & s_{j-1} & s_j & s_{j+1} & \cdots & \cdots \\
\downarrow \quad \nearrow & \downarrow \quad \nearrow & \downarrow \quad \nearrow & \cdots \quad \nearrow & & \cdots \\
r_{j-2} \;\rightarrow & r_{j-1} \;\rightarrow & r_j \;\rightarrow & r_{j+1} \;\rightarrow & \cdots & \\
\downarrow \quad \nearrow & \downarrow \quad \nearrow & \downarrow \quad \nearrow & \cdots \quad \nearrow & \cdots & \\
er_{j-2} \;\rightarrow & er_{j-1} \;\rightarrow & er_j \;\rightarrow & er_{j+1} \;\rightarrow & er_{j+2} & \rightarrow
\end{pmatrix}
$$

The contingent process is, for example, one which is postulated to arise when in a vigilance task attention transiently fails, so that $s_j = null$, but responding continues. An example of such an experiment has been analysed by Gregson (2001).

Transitions between the two processes can arise at any j and each process can be regarded as a state of the system. Then, from the perspective of either **U** or **C**, the system as a whole is nonlinear and nonstationary, but it can be written as a 2-state Markov and, at that level of analysis, can be stationary and stochastic, that is:

$$
\mathbf{T}_{2 \times 2} := \begin{pmatrix} \mathbf{U} \Rightarrow \mathbf{U} & \mathbf{U} \Rightarrow \mathbf{C} \\ \mathbf{C} \Rightarrow \mathbf{U} & \mathbf{C} \Rightarrow \mathbf{C} \end{pmatrix}
$$

For example, a series might exist, a part of which is observed, such as

$$\rightarrow ..., \mathbf{U}, \mathbf{U}, \mathbf{U}, \mathbf{U}, \mathbf{C}, \mathbf{C}, \mathbf{U}, \mathbf{U}, \mathbf{U}, \mathbf{C}, \mathbf{U}, \mathbf{U}, \mathbf{U}, ... \rightarrow$$

and if the **C** epochs are taken as evidence of intermittent malfunction, either cognitive or clinical, then the lengths of runs of **C** are of interest.

In terms of fast/slow dynamics, the processes which generate responses r, er within a trial are fast and unobserved, and the transitions from trial to trial $j \rightarrow (j+1)$ are slow and externally observable.

Another extension of the **U** is made by some theorists, where the distal and proximal stimuli are distinguished. This case arises in the study of the perceptual constancies, such as for size-distance object constancy. Let us, in that case, use D and P as labels to distinguish the two sorts of stimulus, and the mapping from D to P is then physiological when D refers to the physical object's dimensions (some distance away) and P refers to the retinal image of the object as viewed by the observer. The second mapping from Ps_j to r_j is then taken to be psychological, the correction that makes the correlation of r, Ds greater than for r, Ps is assumed to take place at this stage. Rewriting the influence diagram U as uncontingent but distinguishing P and D we now have

$$
\mathbf{U} := \begin{pmatrix}
past & & past & & \mathbf{now} & & future & & future \\
Ds_{j-2} & \rightarrow & Ds_{j-1} & \rightarrow & Ds_j & \rightarrow & Ds_{j+1} & \rightarrow & Ds_{j+2} \\
\phi\downarrow & \phi\searrow & \phi\downarrow & \phi\searrow & \phi\downarrow & \phi\searrow & \downarrow & & \downarrow \\
Ps_{j-2} & \phi\rightarrow & Ps_{j-1} & \phi\rightarrow & Ps_j & \phi\rightarrow & \cdots & \cdots & \cdots \\
\psi\downarrow & \psi\searrow & \psi\downarrow & \psi\searrow & \psi\downarrow & \psi\searrow & \cdots & \cdots & \cdots \\
r_{j-2} & \psi\rightarrow & r_{j-1} & \psi\rightarrow & r_j & \psi\rightarrow & [r_{j+1}] & \cdots & \cdots \\
\psi\downarrow & \psi\nearrow & \psi\downarrow & \psi\nearrow & \psi\downarrow & \psi\nearrow & \cdots & \cdots & \cdots \\
er_{j-2} & \psi\rightarrow & er_{j-1} & \psi\rightarrow & er_j & \psi\rightarrow & \cdots & \cdots & \cdots
\end{pmatrix}
$$

and the influence lines are labelled to indicate whether the postulated causal links are predominantly physiological (ϕ) or psychophysical (ψ). It is assumed that the Ds series is not generated by any deterministic rule; the series may be i.i.d. A point to emphasise is that r_{j+1} may be partly determined before any Ds_{j+1} actually occurs. The extensive literature on contrast and assimilation effects in sequences of judgements, following Helson (1964), may be viewed as an attempt to identify the dynamics of situations in U form.

Limits on Identifiability

From only the external record $\{s, r, er\}$ it is not in general possible unambiguously to reconstruct the linkage patterns in either **U** or **C**. This is a serious problem because the linkage patterns are the simplest model of the process dynamics that are available.

The linkages $s_j \rightarrow r_j$ are not linear mappings, and can be represented in fast /slow dynamics as $s_j \rightarrow \Gamma(Y, a, e, \eta) \rightarrow r_j$ (Gregson, 1995), where the evolution of the complex Γ trajectories is fast, if the sequential slow dynamics $j \rightarrow j + 1$ are ignored or treated as second-order. But there are cascades; for example, in $\mathbf{C}$ we have

$$\forall j : K := r_{j-1} \rightarrow \kappa \rightarrow s_j \rightarrow \Gamma(Y, a, e, \eta) \rightarrow r_j \rightarrow \kappa \rightarrow s_{j+1}$$

where κ is affine. Hence the coherence between the two series $\{s\}, \{r\}$ (even given that both are observable) also masks the internal dynamics of the K-cascades. There are also other cascades possible in parallel.

The most readily generated ambiguities in the dynamics rest on triples over a subsequence $j - 2, j - 1, j$. The diagonal (south-east or north-east) links, if present, mean that influence can run forward for two or more steps. Giving the system both memory and coupling between levels s, r, er implies that two or more paths in a triple, such as $s_{j-2} \rightarrow r_{j-1} \rightarrow r_j$, and $s_{j-2} \rightarrow r_{j-2} \rightarrow r_{j-1} \rightarrow r_j$ exist. If the delay times on these two paths are different (only because they have a different number of internal links each with unit delay) then the shortest should dominate if they run in parallel. A consequence of this is that a path over a quad $j - 3, j - 2, j - 1, j$ could be shorter than a path over a triple.

Perhaps surprisingly, there are many experiments in which the stimulus series is not really known, all that is recorded is the series $\{r\}$. The stimuli are degenerated into a set of states, so that on any one trial there is ambiguity within a state, even if the state is identifiable. Defining an experiment as a set of treatment levels mapping onto a potentially continuously-varying response, as is done in a factorial design, is such as example if repeated measures are used within cells. But the number of repeated measures needs to be so great, for system identification, that an ANOVA analysis becomes dubious pertinent.

Most methods for time series dynamics do demand long series and this monograph is mostly about series which are short but informative, even if they are perhaps not regarded as subsamples which might be con-catenated after experimental replication to form a longer stationary series. Long-term correlations cannot be detected if the data are not themselves long-term (Peng, Hausdorff & Goldberger, 1999).

As they have been written, U and C are examples of Event series. How-ever, if the presentation of a stimulus is allowed to be contingent upon the response latency of the subject's response to the previous stimulus, then the series become Events-in-time series.

If we are only interested in response shifts, such as attention coming on (+) or going off (-), then the response is binary (+,-) and the series is a Time-interval series, the intervals are between crossing points, and the unit of measurement of the inter-crossing-point-interval durations is the single inter-event interval. Such series are also known in statistical the-ory as point processes, the response shifts between on and off states, and the intervals in time between successive crossing points between the two states have a distribution in time which may be studied in its own right.

Some peculiarities of psychophysical time series

Time series in psychophysical experiments are, in some respects, quite dif-ferent from those in areas of application such as physiology or economics. The spacing in real time of the events is two-dimensional, from stimulus to stimulus, and may not be constant but a function of events in the R or E series on previous trials. The spacing in time (the response latencies) from stimulus to response within a trial is again variable, and is contingent on previous trials to some degree, but also on the intrinsic psychophysical mapping $S \rightarrow R$.

We have to therefore distinguish two major cases, contingent and non-contingent. The latter is simpler, so we depict it here as a diagram of in-fluence lines. Each such line may, in theory, be modelled as a regression, not necessarily linear, and may or may not be thought of as a stationary process.

It may be helpful to distinguish the contributing parts of the dynamics of the total system as identified, not identified, or unidentifiable. The human participant in the experiment has a memory and forms concepts, which are here labelled expectations, E. This is an extra series that does not have a corresponding part in most models in the physical sciences. From the viewpoint of an external observer the experimenter the stimuli are or can be identified if they are, in fact, created by the experimenter, the responses are identified, the expectations are unidentifiable as they are private information of the participant in the experiment. They can be asked for and this then constitutes another experiment with another series of self-reported expectations, ψE.

Note that the NPD dynamics, running down the page, are orthogonal to the sequential stimulus dynamics (controlled by Φ) running across the page. It is thus usual in psychophysical experiments to try and space the stimuli in real time so that the Γ mapping becomes independent on each trial, and then averaging over trials under the assumption of stationarity models the form of Γ, which has been known for about 150 years to be roughly ogival. An extra variable, e, called here for convenience *sensitivity*, modifies the shape of the ogive. This spacing of the S sequence can often fail and when it succeeds it destroys information about the dynamics of the total system.

We may remark that all the influence lines are causal, even if not identified, the whole system is deterministic and there are no random parts, though residual observational error on S and R may be treated as stochastic. Some of the nonlinearity in NPD can have the appearance of noise if an attempt is made to model the $S \mapsto R$ relations with a linear model.

There are other important distinctions that are intrinsic to psychophysiological processes. The Φ sequence is in slow dynamics, the orthogonal Γ process is in fast/slow dynamics (as is characteristic of most physiologically-based sensory processes), and the $E_{j-1} \mapsto R_j$ is indeterminate in this respect.

Figure 1.3

Influence Lines Diagram

The non-contingent case, S is not a function of previous R

S = stimulus, R = response, E = expectation

Φ is the stimulus generating function in time

Γ is the NPD (nonlinear psychophysical dynamics) function

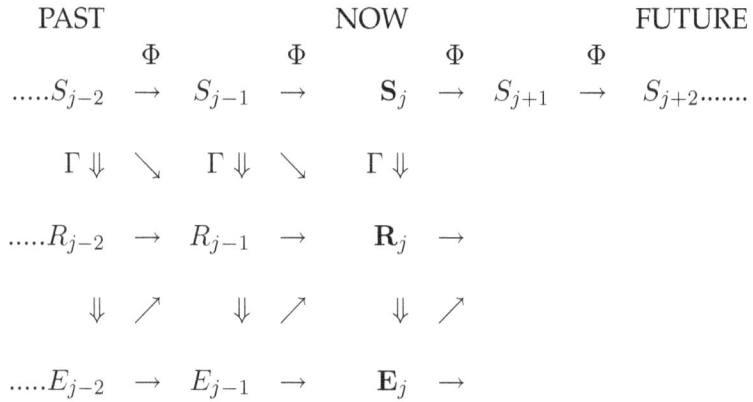

PAST   NOW   FUTURE
$$\begin{array}{ccccccccc}
 & \Phi & & \Phi & & \Phi & & \Phi & \\
.....S_{j-2} & \rightarrow & S_{j-1} & \rightarrow & \mathbf{S}_j & \rightarrow & S_{j+1} & \rightarrow & S_{j+2}....... \\
\Gamma \Downarrow & \searrow & \Gamma \Downarrow & \searrow & \Gamma \Downarrow & & & & \\
.....R_{j-2} & \rightarrow & R_{j-1} & \rightarrow & \mathbf{R}_j & \rightarrow & & & \\
\Downarrow & \nearrow & \Downarrow & \nearrow & \Downarrow & \nearrow & & & \\
.....E_{j-2} & \rightarrow & E_{j-1} & \rightarrow & \mathbf{E}_j & \rightarrow & & & \\
\end{array}$$

$$\Gamma(a, Y(Re), e) \Leftrightarrow \Psi\Phi(S, R, sensitivity)$$

Measure Chains

Time series are modelled usually by assuming they are from a continuous function, then employing differential equations, or in discrete time, and then using difference equations. In dynamical parlance, they are functions on the reals R, or on the integers Z. But, since the work of Hilger (1988), it has been possible to consider intermediate cases where the time series has properties that are neither strictly on R nor on Z, but on a time scale T. A time series built on T is called a *measure chain*. The extensions of these ideas to nonlinear functions are reviewed by Bohner and Peterson (2001).

The delta derivative f^Δ for a function f defined on T has the properties

    i) $f^\Delta = f'$ is the usual derivative if $T = R$

    ii) $f^\Delta = \Delta f$ is the usual forward difference operator if $T = Z$.

Various examples of nonlinear functions on measure chains are reviewed by Kaymakçalan, Lakshmikantham and Sivasundaram (1996).

Our interest in these extensions is partly motivated by knowing that if some process is represented as continuous then its functions may not exhibit chaos, but if it is mapped into discrete time then it may have locally chaotic dynamics. Bohner and Peterson (2001) give examples of biological processes that have gaps and jumps, so that their time basis is not regular, yet they still have biological continuity. These are more readily modelled on T chains than on constant-interval series.

Other Special Cases of Transition Matrices

For attempting a bit more coverage of possibilities but not pretending to any completeness, we should mention some special sorts of transition probability matrices that have potential application in real-world psychological processes. Markovian representation of contaminative dynamics is one example.

In Table 1.1 the terms with a negative sign mean that the process threat returns to the previous state. This somewhat odd and non-standard notation is due to Wang and Yang (1992) and it is, in my view, preferable to rewrite the matrix Q_N as in Table 1.2, which leads to immediate computability from raw sample data. We then have:

Table 1.1: Birth and Death Process of Order $(N + 1)$

$$Q_N = \begin{pmatrix} -b_0 & b_0 & 0 & ...0 & 0 & 0 \\ a_1 & -(a_1 + b_1) & b_1 & ...0 & 0 & 0 \\ ... & ... & ... & ... & ... & ... \\ 0 & 0 & 0 & ...a_{N-1} & -(a_{N-1} + b_{N-1}) & b_{N-1} \\ 0 & 0 & 0 & ...0 & a_N + b_N & -(a_N + b_N) \end{pmatrix}$$

In this process an individual enters at the first state s_0 and stays there over time with probability $p(1 - b_0)$; if, however, he/she becomes infected this then occurs with probability $p(b_0)$ and progression subsequently is through the states in strict sequence to eventual death. In each state $s \in \{Q\}$ the process may stagnate. The penultimate state is thus recovery without reinfection.

If the whole process is to represent persistent reinfection of a subsample of the population then additional off-diagonal terms to create a feedback path to s_1 have to be introduced. Alternatively, a whole matrix for each cohort of infection is written and the population as a whole treated

---

Table 1.2: B and D process in probability notation

$$Q_{p,N} =$$

$$\begin{pmatrix} p(1 - b_0) & p(b_0) & ...0 & 0 & 0 \\ ... & ... & ... & ... & ... \\ 0 & 0 & ...p(a_{N-1}) & (1 - p(a_{N-1} + b_{N-1})) & p(b_{N-1}) \\ 0 & 0 & ...0 & p(a_N + b_N) & p(1 - (a_N + b_N)) \end{pmatrix}$$

---

as a collection of sub-populations running in parallel but staggered in the times of their onsets in the first state s_0. The assumption is that the sub-populations are closed and cannot cross-infect. That might be a plausible way to proceed for a collection of small country towns, separated in terms of social and commercial movements between pairs of towns, but would not work in suburbs of a large city.

An alternative model is given by Wang and Yang, which is shown in Table 1.3, but this does not seem to offer any immediate advantages. The problem is to derive expressions for the expected duration (and frequency distribution of expected durations) of an individual in any state s, which is commonly treated as an exponential distribution with generating parameter $p(s_{ii})$, the leading diagonal term in $Q_{p,N}$.

Table 1.3: Canonical structure of epidemial dynamics

$$Q = \begin{pmatrix} -(a_0 + b_0) & b_0 & 0 & \dots & 0 & 0 & 0\dots \\ a_1 & -(a_1 + b_1) & b_1 & \dots & 0 & 0 & 0\dots \\ \dots & \dots & \dots & \dots & \dots & \dots & \dots\dots \\ 0 & 0 & 0 & \dots & a_n & -(a_n + b_n) & b_n \\ \dots & \dots & \dots & \dots & \dots & \dots & \dots \end{pmatrix}$$

where $a_0 > 0, b_0 > 0, a_i > 0, b_i > 0 \quad (i > 0)$.

The next matrix arises from a theory about the sequential generation of stress, conjectured by Jason Mazanov (personal communication, 2004), which was encoded simply in a linkage flow diagram. This diagram would, unfortunately, confuse two time scales: the short-term scale is between-states within-time, and the long-term scale is within- and between-states but between-times. To disentangle this it is necessary to produce two matrices, and to have long-term as a dummy state for the short term matrix to exit into the long-term, and to be re-entered from it. It would, of course, be possible to nest the short-term states within one larger matrix, as a submatrix on the diagonal, but this is clumsy.

Table 1.4: Jason's Theory

The Short-term Matrix; internal to one episode

State	Stressor	interp1	Stress	interp2	SR	Long-term
Stressor	p_{11}	p_{12}	p_{13}	p_{14}	p_{15}	p_{16}
interp1	p_{21}	p_{22}	p_{23}	p_{24}	p_{25}	p_{26}
Stress	p_{31}	p_{32}	p_{33}	p_{34}	p_{35}	p_{36}
interp2	p_{41}	p_{42}	p_{43}	p_{44}	p_{45}	p_{46}
SR	p_{51}	p_{52}	p_{53}	p_{54}	p_{55}	p_{56}
Long-term	p_{61}	p_{62}	p_{63}	p_{64}	p_{65}	p_{66}

The corresponding long-term matrix has the same core of five stress generation states, but with a link back to Short-term substituted for Long-term as the sixth state. The numerical probabilities are, however, different and the existence of absorbing states would be expected to arise.

Let us make trial substitutions in the 6×6 Short-term matrix, as the off-diagonal pattern does resemble slightly the pattern in the epidemial matrices.

Table 1.5: substitution in Table 1.4

$$STM = \begin{pmatrix} .10 & .80 & - & - & - & .10 \\ .08 & .08 & .64 & - & - & .2 \\ .08 & - & .14 & .48 & .10 & .2 \\ .08 & - & .04 & .16 & .32 & .4 \\ p_{51} & - & p_{53} & - & - & .20 \\ p_{61} & - & p_{63} & - & - & .50 \end{pmatrix}$$

The terms left in p_{xy} form in Table 1.5 are suspected to be the manipulable consequences of the enviroment; in short they could be the experimental variables. The inequalities $p_{51}\{> \: or \: <\}p_{53}$ and $p_{61}\{> \: or \: <\}p_{63}$ have some status in that they conjointly define four alternative theories on the persistence of stress induction.

The obvious problem with this model is that it has 60 d.f., and for system identification at least 1200 d.f. would be needed in the raw data being modelled. There are precedents in econometrics and in epidemiology for models of such complexity, but the identification problems as soon (as feedback loops are nested) become well outside the usual resources of clinical psychology. If the model were to be reduced to being expressed only in externally observable variables, which are Stressors and SR, then there are 8 d.f., which is tractable. It might then be possible to test whether the process is, in fact, reducible to a two-level Markovian dynamic.

Chapter 2

Information, Entropy and Transmission

Until the present, most of our understanding of biological systems has been delimited by phenomenological descriptions guided by statistical results. Linear models with little consideration of underlying specifics have tended to inform such processes. What is more frustrating has been the failure of such models to explain transitional, and apparently aperiodic changes of observed records.

(Zbilut, 2004, p.4)

This monograph is mostly about data that can be characteristically intractable in the face of viable methods that are developed in other disciplines. For example, in creating methods to study series of earth tremors that could be used to predict earthquakes, multiscale analyses are proposed (Zaliapin, Gabrielov & Keils-Borok, 2004). A series is chopped into short segments, where it is known from extensive data and experience they will be approximately linear trends, up or down, separated by turning points. The segments thus created are a mix of slow and fast fluctuations; the slow fluctuations have large amplitude and are the part approximated by a linear trend, the fast fluctuations are lower amplitude and appear to ride on the slow segments. In terms of filtering, each segment in succession is separately filtered, and a model that involves both the parameters of each segment and the locations of turning points can be constructed. The literature in probability theory on identifying and predicting such processes is now diverse and not restricted to variations of linear models. The interrelations of symbolic dynamics, Markov chains, discontinuities or singularities, and random evolutions feature in modern mathematical treatments (Horbacz, 2004), and will receive some usage here.

The important contrast with a model in experimental psychology is that earthquakes are, unfortunately, not controlled, they are observed, whereas stimulus-response relationships are manipulated, they are explored in the laboratory where conditions are set up to create some sort of local stationarity. This often makes it possible to use models like the so-called Stevens' Law,

$$Resp = a + Stim^b + \epsilon$$

where a and b are scalars, and $\epsilon \sim N(0, \sigma)$ is Gaussian noise, the Stimuli can be scaled in some metric based on physical properties and the Response may be arbitrarily scaled or represented as a probability. The problem is that, unless we replace the constants a, b by functions varying over time, there is no way the underlying dynamics that are involved when the system is perturbed and comes back to stability, at some rate, can be captured in the model. Stationarity is censoring of dynamics. Ideally, what we seek is a relaxation of metric assumptions and insight into dynamics at the same time (Gregson, 1988, 1992).

It has been noted that the filter in use is also part of the current model being tested and, more usually, it is the model that is the focus of attention and the structure of the filtering, together with the notion of the statistical nature of the noise, that is taken for granted as a secondary consideration. An example is the practice of multiple regression, where the data are either static outside time, or time series. The residual noise after model fitting is taken as Gaussian i.i.d. and independent of the signal components.

Here in regression we have a multiplicity of predictor variables and one dependent variable, so the input to the filter is a set of parallel time series and the problem is to find the intervariable causality of the system and filter optimally to exclude noise. This problem is usually treated as hypothesis testing in psychometric literature, but a strong alternative approach, which can be grounded in Bayesian inference, has been advanced by Burnham and Anderson (2002), which uses instead the corrected likelihood inference approach built on Akaike's Information Criterion (AIC) and its many modifications. There are still difficulties with doing this for small samples with many alternative models, and it is preferable theoretically to restrict consideration to nested models. There are other criticisms of hypothesis testing in the specific context of identifying the dynamics of causality that have been pointed out by Feng (2003).

A little-known theorem in what is called the search problem, another name for the inference or inverse problem (which means getting back from data to an optimal causal model), was derived by Wolfert and Macready (1997). This states that, as no model is ideal, uncontaminated, pure truth, all algorithms that address the inverse problem are equally bad, but even so if the search algorithm is fitted to a particular problem space, it can work better than most other alternatives that do not use all the prior information that properly define the inverse problem. It does not follow from this that all filters are equally bad but, as remarked in the introductory chapter, false outcome information about causality can be created by inefficient filtering.

Consider a case where there are potentially 5 predictor variables $i = 1, ..., 5$, so there are $2^5 - 1$ possible models, ignoring the situation where nothing has any predictive value. The linear models are

$$y = b_1 v_1 + b_2 v_2 + b_3 v_3 + b_4 v_4 + b_5 v_5 + \epsilon \qquad [2.1]$$

and all subsets of this generic model where some b_i are zero. Each operation of setting one or more $b_i = 0$ is a filter, apart from the filter implicit in partitioning the residual ϵ. The models are said to be nested if they are ordered from [2.1] down to $y = b_1 v_1$, dropping terms v_5, then v_4, then v_3, and so on. This is simply making the filter progressively more stringent on an implicit ordering of the potential relative importance of the variables to carry information about y. There are still 5!=120 nested models to consider. If cross-product terms on the v are included, making the models nonlinear, then the number of possible models explodes and nesting them is, in reasonable, practical terms, computationally intractable. Any overdetermined model will fit data in the sense of creating some prediction of input-output relationships; any overdetermined model is an approximation to a transparent filter.

Shadowing

The concept of shadowing is fundamental to describing how a filter that is only approximately a representation of the mathematics that defines the generation of a time series of data can be modelled, under restricted conditions, using information measures. The explanation rests on three stages in the argument: what is meant by noise, what is meant by a nearby trajectory, and why moving into symbolic dynamics, with Markov representations, is a constructive approach to identifying and controlling nonlinear dynamical trajectories. In short, shadowing is a type of filtering to preserve stability.

For linear time series, the noise is assumed to be in some frequency band that separates it from the signal or, if a frequency domain partition is not possible and Fourier analyses are invalidated, then signals are assumed to be the sharp as opposed to the diffuse components. This, again, will not work if the signals are not at least in part periodic. It is necessary to distinguish between two sorts of noise: that intrinsic to nonlinear dynamical evolution, and that due to added error of observation (Grassberger, Hegger, Kantz, Schaffrath and Schreiber, 1992).

Consider a series of symbols $Y_1, ..., Y_j, ..., Y_N$, that is where $Y \in \{Y_a, Y_b, Y_c, .., Y_S\}$ takes at any time one of S exhaustive mutually exclusive states. In the limit if S is large and the states are ordered the symbols Y are replaced by some variable y_j. The recursive difference equation of the system is then

$$Y_{j+1} = f(Y_j) \tag{2.2}$$

and real data will be of the form

$$y_{j+1} = f * (y_j) + \epsilon \tag{2.3}$$

where f and $f*$ are in some definable sense close, and where ϵ is experimental observation noise, distributed hopefully over a closed interval.
Such noise can be additive, as written or multiplicative. Noise with infinite
variance is obviously troublesome but exists in some statistical models.
The other sort of noise is added by the dynamics during its evolution and
it is asked if there exists a nearby trajectory that is free from this noise but
also satisfies the exact dynamics. Such a trajectory does not have to exist,
but if it does, it is said to be **shadowing** the exact dynamics, always within
some small limit close to the desired clean model. Shadowing is thus a sort
of filter, finding it is an art. As there are two sorts of noise, if an attempt
is made to filter out ϵ, then the other pseudonoise gets partly taken out
with it and distortion of the identification of $f(y)$ results. The question is
whether misidentification of $f(Y)$ also follows. It seems that, under some
technical conditions, when a representation by Markov chains is possible (that is, using Y and not y) it is possible to construct shadowing even
when the dynamics involve singularities (Krüger and Troubetskoy, 1992).
The mathematical basis that supports shadowing is the Anosov Closing
Lemma (Katok and Hasselblatt, 1995, section 6.4.15). It is easier to meet
shadowing conditions if the dynamics involve some cycling, which in turn
implies a particular structure in a Markov chain.

Turning the argument around, so that we go from y to Y and not Y to
y, any series y can be converted to a Y series, by rescaling onto the unit
interval, that is

$$\langle \max(y), \min(y) \rangle \Leftrightarrow \langle 1, 0 \rangle =_{def} \langle \max(Y), \min(Y) \rangle \tag{2.4}$$

and then the Y representation can be partitioned into S subsets. These
subsets are the states of the system in symbolic dynamics terms, but only

if the probabilities of state occupancy are equal, that is

$$\forall s_i \in S, \ p_i = 1/S$$

is the partitioning maximum entropy. Otherwise, the widths of the s_i have to be adjusted to make the p_i approximately equal, so that the filter implicit in the partitioning transmits the most information.

The interrelations between symbolic dynamics, Markov chains, shadowing, and recurrent neural networks have an extensive and expanding literature (Lawrence, Tsoi and Giles, 1996; Setiono & Liu, 1996; Zak, 2004), but so far this has had little impact on psychometrics, though some on neurophysiology.

Fitting a Model plus Filter

The problems with fitting a family of models are various, the most troublesome is that of overfitting. In short, any model that is too complicated will fit data, as a filter it becomes transparent, the flap lets the dog through as well as the cat. If the process is really nonlinear and not stationary, then trying to capture those dynamics economically by adding the product terms such as $b_{1,2}v_1v_2, b_{1,2,3}v_1v_2v_3$, of which there are at least $^5C_2 +^5 C_3 +^5 C_4$, creates a lot of models where the fit assessed as R^2 will be almost equally high and "significant" but useless. Burnham and Anderson (2002, chapter 3) give a valuable set of examples of fitting models by AIC to various data sets; the nearest to our present interests is the example of adding an insecticide to a tank full of earth in the bottom and plants and little fishes, to simulate a pond. The example involves dynamics changing over time and illustrates the point that various analytical methods need to be employed in parallel and that we have to make maximum use of what is already known about the problem in terms of its basic science, to select a closed set of candidate models to achieve parsimony and relevance.

Working in discrete and not continuous time and considering a variable $y_j, j = 1, .., N$ that takes only a finite set of n symbols, each symbol

$s_i, i = 1, .., n$ in a time series sample has probability p_i, then the information is

$$H(x) = - \sum_{1}^{n} p_i log(p_i) \qquad [2.5]$$

The Kullback-Leibler information that compares a model θ, giving corresponding expected probabilities $x_i | \theta$ with a data sample is then

$$I(y, x) = \sum_{i=1}^{n} p_i \cdot log \left(\frac{p_i}{x_i | \theta} \right) \qquad [2.6]$$

and that may be thought of as a mismatch distance (not a true metric distance, because of asymmetry) between theory and data, in our situation θ becomes a filter and the mismatch is expressed in information loss. Then if there is a class of related and a priori plausible parsimonious models θ_1, θ_2, etc., the models are ordered in terms of their mismatches, and iterative revision to make the filter open, as the model converges on data, is by revision of θ evaluated by minimising mismatching. The approach using Kullback-Leibler information has been extended to the situation where we do not know how the data are generated but can consider three hypotheses at the same time, one of which is some sort of noise (Zheng, Freidlin and Gastwirth, 2004).

If the models considered are linear and fitted to short stationary time series, and we are only concerned with one data set n, then AIC can be used to select a (or a subset) most appropriate candidate model. For a model θ, with r parameters and residual sum of squares (RSS)

$$\hat{\sigma}^2 = RSS/(n - (r + 1))$$

for least squares estimation with log likelihood

$$log(\Lambda(\hat{\theta})) = -\frac{1}{2} nlog(\hat{\sigma}^2) - \frac{n}{2} log(2\pi) - \frac{n}{2} \qquad [2.7]$$

and we can drop the last two terms as they do not affect inference because they are constants for any competing model in a closed set on one data sample. Then for a given data set y

$$AIC = -2log(\Lambda(\hat{\theta}|y)) + 2K \qquad\qquad [2.8]$$

where K is the number of fitting parameters ($\geq n$). The AIC formula as just given has limitations and has been modified in various ways for small samples. A very important recent result (summarised by Burnham and Anderson, 2002, section 6.4.5) is that a set of weighted AIC values for a set of models on one data set yields a good approximation to the Bayesian posterior model probabilities, provided that one is prepared to assume equal prior model probabilities.

As the flow diagram of filtering in Chapter 1 is drawn, these AIC methods could be applied where we have linear models to consider on the stationary subsets used for filter revision, but not globally. As exploration and tests of non-stationarity are the main focus of some later chapters here, the AIC results are given for completeness but not for invariant employment. They are still preferable to hypothesis testing and significance measures in many situations.

Preliminary Data Examination

Any time series is an ordered set of symbols. Those symbols may each have numerical properties or they may not. If they are numbers in a metric, they may be re-encoded as weaker numbers with only ordinal properties, they may be re-encoded as ordered non-numerical symbols, or they may simply be labelled as elements of a mutually exclusive exhaustive set of states. What sort of symbols they are imposes limits on what meaningful operations may be performed on them. If it is assumed that the numbers are in a mathematical sense well-behaved, and they are not, then bogus conclusions could be drawn from operating on them algebraically.

Given a time series, a finite sample from some longer process, there are a wide range of other series that may be derived from it. If we do not know the history of the data with which we are presented, then it may be that what is observed is itself a derived series. Two successive numbers,

y_j, y_{j+1}, observed at times t_j, t_{j+1} respectively, may be replaced in various ways. The simplest are the difference $\delta_j y = y_j - y_{j+1}$, and the absolute difference $|y_j - y_{j+1}|$. This operation of differencing may be repeated to a kth order. If the difference $\delta_j t = t_j - t_{j+1}$ is not constant, then the local gradient $g_j = \delta_j y / \delta_j t$ may be used to create a new series of g values. The statistical properties of derived series are not in general the same as those of the core series from which they are derived. Each of the series $y, \delta y, g$ may have computable first and second moments of the distribution of values that their variable takes.

Relaxation of Metric Assumptions

Serious difficulties are met in identifying underlying dynamical processes when real data series are relatively short and the stochastic part is treated as noise (Aguirre & Billings, 1995), it is not necessarily the case that treating noise as additive and linearly superimposed is generically valid (Bethet, Petrossian, Residori, Roman & Fauve, 2003). Though diverse methods are successfully in use in analysing the typical data of some disciplines, as in engineering, there are still apparently irresolvable intractabilities in exploring the biological sciences (particularly including psychology), and a proliferation of tentative modifications and computational devices have thus been proposed in the current literature.

The theoretical literature is dominated by examples from physics, such as considerations of quantum chaos, which are not demonstrably relevant for our purposes here. Special models are also created in economics, but macroeconomics is theoretically far removed from most viable models in psychophysiology. Models of individual choice and the microeconomics of investor decisions may have some interest for cognitive science, but the latter appears to be more fashionably grounded, at present, in neural networks, though again the problem of simultaneous small sample sizes, nonlinearity, non-stationarity and high noise have been recognised and addressed (Lawrence, Tsoi & Giles, 1996).

Much of the computational literature focusses on fractal dimensionality, Lyapunov exponents, or entropy (Mayer-Kress, 1986), however, there are paths between symbolic dynamics and entropy measures, particularly where the location of periodic saddle orbits is involved (Lathrop and Kostelich, 1989a,b). The use of Lyapunov exponents emphasises that the predictability and controllability of processes are fundamental concerns, but even here systems can float between uncertainty and certainty about their future evolution (Ziehmann, Smith & Kurths, 2000). One exception, that could circumvent the difficulties in analysing psychological data that are encoded in numbers that do not satisfy metric axioms, is to use symbolic dynamics. The problem of constructing psychological measures with axiomatic bases that define some sort of metric was a continuing challenge in the late-20th century (Krantz, Luce, Suppes & Tversky, 1971) and the invalidity of assumptions by fiat that had been achieved has been described by Michell (2002). Possible foundations of psychophysical scaling were established on a basis of the long-established functional calculus (Aczél, 1966, gives an historical survey) to define what forms of scales can satisfy metric axioms (Luce, Bush & Galanter, 1963). The assumptions therein imply functional deterministic stability and are not of use for modelling dynamics processes without augmentation; the modern approach using MCMC statistics can in some restricted cases be treated as a hybrid of Euler functionals and Markov chain transitions (Winkler, 2003, p. 314), which does postulate an evolving mixture of deterministic and stochastic processes running through time. The symbolic dynamics explored here are closely related to some of the assumptions of MCMC practice, but we do not use the full apparatus of statistical estimation, rather the focus is on the ubiquitous non-stationarity of psychological time series. In effect, by using symbolic dynamics, the processes studied are taken to be in the class of discrete dynamical systems, and tests for stability that are developed are in that domain (Gumowski & Mira, 1980).

There have apparently been examples from social psychology where using symbolic dynamics instead of making metric assumptions (with ANOVA-type statistics) has produced more sensitive insights into the dynamics (Heath, 2000, p. 311). Guastello, Hyde and Odak (1998) and Guastello (2000) on information exchanges during creative problem solving in a social group, Guastello, Nielson and Ross (2002) in the analysis of brain activity in MNR pattern sampling, and Pincus (2001) in family interaction dynamics, have also made valuable use of symbolic dynamics, employing some necessary variations in technical details.

If one wishes to explore nonlinear dynamics directly within the traditional framework of difference-differential equations then the use of variables with metric properties, such as can more readily be achieved in psychophysiology than in psychophysics, is mandatory.

The matrices that we create for a representation of psychophysiological and psychophysical time series are Markovian, and are necessarily square and non-negative. They may also be sparsely filled and quasi-cyclic. We know from the fundamental mathematics and from examples constructed (Mitchener and Nowak, 2004), that if we accept the positivity of the largest Lyapunov exponents as a sufficient indication of chaos, of one sort, then the processes being represented in transition probability matrix form may be chaotic.

One of the powerful consequences of using symbolic dynamics on maps on the unit interval is that Markov transition probability matrices may be created, and from those an information theory treatment is supported, leading back into entropy calculations. The deep mathematical relations between symbolic dynamics, Markov chains, and entropy measures have now an extensive literature, which has been surveyed by Blanchard, Maas and Noguiera (2000).

This rests on some theorems of Parry (1964, 1966) showing that, if a series behaves locally like a Markov process, then, from the perspective of information theory, it is Markov. This approach has been extended by Buljan & Paar (2002).

Symbolic dynamics have also been linked with nonlinear psychophysics (Geake & Gregson, 1999), for encoding the existence of embedded recurrent episodes within trajectories. They are used in the generation of the entropic analogue of the Schwarzian derivative, ESf, in scaling quasiperiodic psychological series (Gregson, 2002; Gregson & Leahan, 2003).

The idea of employing Markovian representations of psychological processes evolving through time is certainly not an innovation and was used in a fundamentally different way in learning theory (Bower & Theios, 1964). There, the dependent variable was the probability of making a particular response (usually a correct one) on a trial in a learning curve that eventually entered an error-free absorbing state; a major finding was that three theoretical states in discrete time generated closer fits to observed data. Mathematical learning theory was not conceptualised as an instance of nonlinear dynamics, but rather as simpler stochastic processes with associated statistical tests.

Parry usefully distinguishes between two matrices that play a part on theory, the State Transition Matrix (s.t.m.) which Parry calls the Structure Matrix, where each cell is 0, or 1 only if a transition exists, and the Transition Probability Matrix (t.p.m.) where each cell is a probability. As the process is Markov, the transition is taken to be $t \to t + 1$, in real time the increment is t to $t + \theta$, where θ is the time interval between two successive observations of the process. In simpler treatments θ is taken as a constant, but it can be a random variable. Given a set S of states $s \in \{1, .., i, j, .., n\}$ that is exhaustive, but not necessarily ordered in terms of some measure $\mu(s)$, the elements of t.p.m, $t(i, j)$ are given by

$$t(i, j) = \begin{cases} > 0 & \text{if } s(i, j) = 1; \\ = 0 & \text{if } s(i, j) = 0 \end{cases} \qquad [2.9]$$

The t.p.m matrices are usually taken in terms of succession, so that each $t(i, j) \in F$ is the probability of state j following state i, $t \to t + 1$. This is appropriate in a dissipative and irreversible process, such as a real psychological time series. But the reverse matrix can be computed, in which each $t(i, j) \in R$ mean the probability that i is preceded by j, $t \to t - 1$. This usually has different eigenvectors. Possibly the most important extension of these ideas is to non-homogeneous Markov chains as a structure for non-stationary psychophysics.

As we will want to examine some real and theoretical Markovian matrices and their associated eigenvalues, it is proper to begin by restating the Perron-Frobenius theorem (Frobenius, 1912, Seneta, 1973) for primitive matrices:

Suppose that T is an $n \times n$ non-negative primitive matrix. Then there exists an eigenvalue r such that:

(a) r real, > 0;

(b) with r can be associated strictly positive left and right eigenvectors;

(c) $r > |\lambda|$ for any eigenvalue $\lambda \neq r$;

(d) the eigenvalues associated with e are unique to constant multiples;

(e) if $0 \leq B \leq T$ and β is an eigenvalue of B, then $|\beta| \leq r$, and $|\beta| = r$ impies $B = T$.

(f) r r is a root of the characteristic equation of T.

We are also going to need to make reference to the well-known scrambling property of Markov matrices, namely

An $n \times n$ stochastic matrix $p + \{p_{ij}\}$ is called a *scrambling* matrix, if given any two rows α and β there is at least one column, γ, such that $p_{\alpha\gamma} > 0$ and $p_{\beta\gamma} > 0$.

A corollary follows; if $Q = \{q_{ij}\}$ is another stochastic matrix, then for any Q, QP is scrambling, for fixed scrambling P.

Fast/Slow Dynamics

Interest in processes where there is a functional division between slow dynamics, that may serve as a carrier, and fast dynamics, that can resemble noise or a signal with chaotic characteristics, possibly began in engineering, but this now recognised as a paradigm for some biological applications. In fact this area of investigation is intimately linked to the two-attractor problem just discussed above. Arecchi (1987, p.42) from a frequency domain approach, observed:

We have shown that, whenever in nonlinear dynamics more than one attractor is present, there are two distinct power spectra:

i) a high frequency one, corresponding to the decay of correlations within one attractor;

ii) a low frequency one, corresponding to noise induced jumps.

The term fast/slow is used to label such processes[1]; it is not, in fact, critical whether the fast part is treated as the signal or the slow part; the important consideration is which of the two parts, if they are separable, is externally controllable over some finite time interval, at some rate of intervention.

For a very simple case, we assume that the dynamics are stationary and first construct the s.t.m. of the slow part, which we label C_{ss}. The double suffix is r to remind us that the process is assumed to be slow and stationary.

$$C_{ss} = \begin{pmatrix} 0 & 1 & 0 & 0 & \dots & \dots & 0 & 0 & L \\ 1 & 0 & 1 & 0 & \dots & \dots & 0 & 0 & 0 \\ 0 & 1 & 0 & 1 & \dots & \dots & 0 & 0 & 0 \\ 0 & 0 & 1 & 0 & \dots & \dots & 0 & 0 & 0 \\ \dots & \dots & \dots & \dots & \dots & \dots & \dots & \dots & \dots \\ \dots & \dots & \dots & \dots & \dots & \dots & \dots & \dots & \dots \\ 0 & 0 & 0 & 0 & \dots & \dots & 0 & 1 & 0 \\ 0 & 0 & 0 & 0 & \dots & \dots & 1 & 0 & 1 \\ L & 0 & 0 & 0 & \dots & \dots & 0 & 1 & 0 \end{pmatrix}.$$

This matrix is defined over transitions in the real time interval $t, t + \theta$, and θ has to be chosen by trial and error if the generator of the slow dynamics (such as a sinusoid) is not known. If θ is too small then some terms

---

[1] Guckenheimer (2003) uses the term slow-fast in a more complicated treatment of the bifurcations of such systems.

$s(i, i) \simeq 1$, and if θ is too large then other off-diagonal cells are not zero. C_{ss} may then be part of a matrix where the elements of S are strictly ordered and the cell $t(..)$ values fall off monotonically as we move away from the leading diagonal.

Let any one line (with correction for the end lines) of the t.s.m. of C_{ss} contain the terms $t(i, i - 1), 0, t(i, i + 1)$, and the minima over S be $\min(t(i, i - 1)), \min(t(i, i + 1))$. This double off-diagonal matrix with the minima substituted for all i is T_{ss}.

If the trajectory is on a closed orbit, then the cells $s(i, n)$ and $s(n, i)$, marked L in the matrix, are also non-zero. The matrix is then a circumplex (Shye, 1978). This form corresponds to an attractor on a limit cycle. Obviously, both with and without the L cells the C_{ss} pattern depends on finding an order of the elements of S that generates the pattern. If the n states are on a closed orbit, then there are 2n such orderings: one can start at any state and go in either direction round the orbit.

If the fast component is the trajectory of a chaotic attractor in its basin, then it eventually visits everywhere, has no absorbing states, and its t.s.m. is ergodic. Call this matrix D_{ns}. The observed matrix of the process is $M_{s+n,s}$ and if the two parts add linearly within any cell then, over some subsequence in time, that is stationary,

$$M_{s+n,s} = T_{ss} + D_{ns} \qquad [2.10]$$

so subtracting cell-by-cell

$$\hat{D}_{ns} = M_{s+n,s} - T_{ss} \qquad [2.11]$$

It is $\hat{D}_{ns}$ that we now treat by Parry measure to find the approximate eigenvalues of the fast part of the system.

The matrices C_{ss} and F from some depression data (Gregson, 2005) do have some interesting resemblance to a symbolic dynamics representation of the Belousov-Zhabotinskii chemical reaction described by Lathrop and Kostelich (1989b, p. 152). *The resemblance does not reduce to saying that the causality is the same between chemistry and psychophysiology, it says merely that a problem in identifiability has common structure.* This is

$$BZ = \begin{pmatrix} 0 & 1 & 0 & 0 & 0 & 0 & 0 \\ 0 & 0 & 1 & 0 & 0 & 0 & 0 \\ 0 & 0 & 0 & 1 & 0 & 0 & 0 \\ 0 & 0 & 0 & 0 & 1 & 0 & 0 \\ 0 & 0 & 0 & 0 & 0 & 1 & 0 \\ 0 & 0 & 0 & 0 & 0 & 1 & 1 \\ 1 & 0 & 0 & 0 & 0 & 1 & 0 \end{pmatrix}.$$

This type of matrix can be evidence of cyclic stable dynamics, on a countable space chain; a deeper analysis is given by Meyn and Tweedie (1993, p. 115). The off-diagonal array and the recurrence at $s(1,7)$ resemble half of the symmetry of Css, and thus the weak asymmetry of F. The BZ dynamics are associated with closed orbits and an intermittent burst after which the dynamics return to a periodic orbit. It is possible to employ the symbolic representation to calculate the topological entropy h_t using only relatively slow orbits and hence of low period, as Lathrop and Kostelich (1989) also had the maps of the reconstructed attractor and the data series 8400 points long.

If N_p is the number of periodic points for the pth iterate of the return map, then

$$h_t = \lim_{p \to \infty} \frac{1}{p} log_2 N_p \qquad [2.12]$$

but [2.12] does not work for short series. Alternatively from the $t(i,j)$ matrix BZ

$$N_p = trace \ M^p \qquad [2.13]$$

The trace[2] is the sum of the diagonal elements of the t.p.m. matrix, as it is dominated by the largest eigenvalue λ_1 of the matrix for large p,

$$h_t = log_2 \lambda_1 \qquad [2.14]$$

which resembles the Parry measure derivation. In the example of BZ, $h_t = 0.73 \ bits/orbit$.

---

[2] See Horst (1963) for an introduction to trace properties and computation.

In situations where there are two or more attractors and fast/slow dynamics exist, a link emerges between the two attractor situation examined above and also, apparently, with nonlinear psychophysics (Gregson, 1988). This is illustrated in an example by Arecchi, Badii and Politi (1984) who show that, if there are jumps between basins of independent attractors, the system as a whole exhibits a low-frequency component in its power spectra. The Lyapunov exponent is then complex, being made up of parts corresponding to attraction and repulsion. This has some qualitative parallel with the Γ function used in nonlinear psychophysics; both Γ and the example created by Arecchi *et al* are grounded in a cubic map, without noise. The dynamics are very complicated and cannot be reduced to a single $1/f^b$, $\frac{1}{2} < b < 2$, power spectrum.

Filtering Sequential Dynamics

The information in a dynamic series involves not just the values taken in a raw data series y_j, but their successive differences, $\Delta^1 y_j, \Delta^2 y_j$ and so on until they converge to a very small value or zero.

Entropic Analogue of the Schwarzian Derivative.

This method was introduced by Gregson (2001), it is a parallel of a derivative introduced by Schwarz (1868), but based on local summations of coarsely scaled series and not on point derivatives of a continuous function. The idea of treating a dynamical trajectory from an entropy perspective is not novel, but is well developed (See Sinai, 2000, Chapter 3) and can be traced back to statistical mechanics in the 19th century.

A series of a real variable y is partitioned into k exhaustive and mutually exclusive subranges in the values y it takes, $k = 10$ is initially sufficient. As a condition of the normalisation $0 \leq y \leq 1$ the end subranges $h = 1, k$ will initially not be empty. Some or all of the remaining $k - 2$ subranges may be empty, when the dynamics are minimally informative. Call the width of one such subrange $\delta^{(0)}$ (of y). $\delta^{(0)} = .1$ for the original y but will almost always be less for $\Delta^m(y)$, the successive m^{th} absolute differences of the series. This δ is the *partitioning constant* of the system. Call the partitioning constant of the range of the mth differences $\delta^{(m)}$. In the examples used here $\delta^{(1)}$ is referred to simply as δ in the tables of results, and all $\delta^{(m)}, m > 0$ are set constant $= \delta^{(1)}$.

The frequencies of observations lying in one segment $\delta_h, h = 1, ..., k$ is then n_h, and converting to probabilities p_h we compute the information in that subrange. The absolute differences of the rescaled y series are taken putting

$$\Delta^1(y_j) = |y_j - y_{j-1}| \qquad [2.15]$$

and further differencing repeats this operation, so that

$$\Delta^2(y_j) = |\Delta^1(y_j) - \Delta^1(y_{j-1})| \qquad [2.16]$$

This operation can be continued only until all absolute differences are zero. Going only as far as Δ^4 is sufficient. Summing over all subranges gives the total information in the mth differenced distribution as

$$I_{\{h\}}^{(m)} = I^{(m)} = \sum_{h=1}^{k} p_h log_2(p_h)|_m \qquad [2.17]$$

Then, by definition, the entropic analogue of the Schwarzian derivative, abbreviated to ESf, has the form

$$ESf := \frac{I^{(3)}}{I^{(1)}} - \frac{3}{2}\left(\frac{I^{(2)}}{I^{(1)}}\right)^2 \qquad [2.18]$$

For some strongly chaotic (theoretical variants on Γ) series, ESf is positive in the examples we have seen and becomes increasingly negative as processes are more stochastic. It has been applied to series from psychophysics, EEGs, climate and economics. Its main advantage is that is is computable over much shorter time series than the Lyapunov exponents.

By extension, each of the distributions of $\Delta^k y_j$ can be predicted by a dynamical model and then compared to data by the Kullback-Leibler formula.

Bispectral Kernel Analysis

This method is usually employed in the frequency domain, but here a time domain version is used, together with the surrogate tests.

Bispectral analyses are, in fact, third-order kernels of time series. They have been extensively used in anaesthesiology in the tracking of the evolution of EEGs during surgery, following Rampil (1998) and Proakis *et al* (1992), and are there computed in the frequency domain using FFTs. They resemble the kernels used in nonlinear analyses described by Marmamelis and Marmarelis (1978).

If there exists a real discrete zero mean third-order stationary process x, then its third-order moment matrix is defined over a range of lags $1, ..., m$, $1, ..., n$ by

$$R(m, n) := E[x(j) \cdot x(j + m) \cdot x(j + n)] \qquad [2.19]$$

This matrix is skew symmetric as m, n can be exchanged in the definition. The triangular supradiagonal matrix is sufficient for exploratory purposes and is used in the tables in the form shown; to compute its eigenvalues, it is reflected into the square form and the leading diagonal cells filled with the average of the off-diagonal cells as an approximation. Its roots will, in general, be a mixture of reals and complex conjugates.

$$
\begin{array}{lllll}
b(1,2) & b(1,3) & b(1,4) & b(1,5) & b(1,6) \\
b(2,3) & b(2,4) & b(2,5) & b(2,6) & \\
b(3,4) & b(3,5) & b(3,6) & & \\
b(4,5) & b(4,6) & & & \\
b(5,6) & & & &
\end{array}
$$

Compare the usual second-order autocorrelation which is defined as

$$R(m) := E[x(j) \cdot x(j + m)] \qquad [2.20]$$

over a range of lags $1, ..., m$.

Chapter 3

Transients onto Attractors

There are various ways of dealing with transient perturbations of what appears to be an almost regular evolution of dynamics: one can ignore them; one can delete them and replace then by the moving average of adjacent observations; one can treat them as outliers, perhaps generated by errors in measurement; one can treat them as outliers from a non-normal distribution due the the presence of a secondary distribution; or one can treat them as intrinsic to the dynamics and perhaps predicted by chaos theory. We will examine some alternatives. In psychometrics, it is tempting to regard them as errors due to the influence of unwanted brief external stimuli. One can argue (Gregson, 1992) that any psychophysical stimulus-response series is itself a series of transient perturbations of the stable dynamics of a sensory system that exist when no inputs are being received. A sensory system then filters signals and returns after each to stability, but may modify its own filter characteristics in so doing. It can show adaptation and desensitisation.

A reanalysis of long-standing assumptions in models of judgements, as made in multidimensional scaling or similarity theory, has shown that some metric space and stochastic assumptions are neither necessary nor universally valid (Miyano, 2001), but, importantly, using a cross-entropy formulation (resembling a Boltzmann or Kullbach-Leibler measure), it is possible to tie together imprecision of judgements and their associated confidence levels, and to have, within the one model, behaviour ranging from stochastic to deterministic. Such floating appears to be a common characteristic of psychophysiological processes at various levels from the neurophysiological to the psychophysical and it is interesting that researchers in various of these areas have resorted to local entropy measures rather than to the more traditional frequency domain representations.

Let us first look at an approach from nonlinear dynamics. The purely mathematical analyses of nonlinear dynamics are concerned with the properties of attractors under stability, within their own basins, or at the boundaries of their basins. It is in applications to the physical sciences that questions of stability have mainly been studied as computational problems (Viana, 2000). The use of nonlinear dynamics in the behavioural sciences is still slight, though advocated in some quarters (Nowak & Vallacher, 1998), and the serious question that investigators face is what to do about the fact that most data are about transients or pathways onto stability that, at any moment, might be overwritten by the effects of new destabilizations. In previous work, I have gone so far (Gregson, 1992) as to equate, conceptually, stimulation to transient destabilization of a continuing underlying process that is itself nonlinear and may be chaotic.

Nonlinear dynamics are analysed as a special class of problems within time series analysis, if we take a statistician's approach. Most of the available methods and the derived indices which characterise the nature of the nonlinearity and the attractors generating the dynamics are legitimate only for long stationary series (Kantz & Schreiber, 1997). There are other problems (Demazure, 2000) in that linearization is not universally applicable, but may be applicable in the neighbourhood of singular points that are attracting. There are also difficulties in linearization when working with dimensionalities greater than 2 and we do not usually know much about the dimensionality of psychological dynamics a priori.

In reopening the questions here, initially a restriction to considering bivariate trajectories in the reals will be imposed. One of the system parameters may be designated *gain*, and the other *control*; that is not strictly necessary, but conceptually it can help. The output variable of the system will be single and real, at least initially. This will be relaxed once some simpler methods have been outlined. A diversity of indices have been proposed for characterising the dynamics which are supposed to underlie the generation of any observed trajectory of sufficient length; these have variously plausible mathematical justifications and, in some cases, may be treated as statistics and not as strictly deterministic indices, so that confidence intervals on estimates may be derived either analytically or by Monte Carlo methods.

Various reasons exist for being interested in very short time series when these are taken to be samples from a dynamical trajectory. We may suspect that there is evidence of moving onto or off an attractor, or behaviour in the neighbourhood of a singularity, or of one of the local recurrent patterns within a longer series that jumps between levels of stability. In the most general sense, we may simply think the process is not stationary, but not be sure how or why this happens. In psychological processes the very nature of the nonstationarity itself can be informative and even characterise the system.

Setting these constraints, the dynamics of the process can be represented on the surface of a manifold which has locally (geodesically) at least two dimensions but is in an embedding space (geometrically) of at least three dimensions (Gregson, 1998; Roweis & Saul, 2000). The manifold of a nonlinear system is not flat, if it is a surface in two dimensions, but has curvature in varying degree in at least one direction corresponding to the axes of the embedding space. In psychometric applications, this manifold may be the response surface of a process which maps from stimulus properties to observable responses. It is thus the extension of the traditional idea of a psychometric curve, but any one curve is produced by a section plane through the manifold, and where the section is drawn is a function of the task constraints imposed in generating a small region of the manifold. Reverting back to the idea of the influence line diagrams U, C in the introduction, a response time series is the result of meandering over the surface of the manifold. The data of an experiment are thus a sparse mapping of the surface of the manifold in this restricted case where the manifold is a surface without volume.

Whether or not the manifold can be plausibly reconstructed from a data set depends not just on the degree of sparseness but on where the sparse data lie; if they are tightly clustered, then a local region of the manifold can be recovered, partly because in that local region geodesics on the surface of the manifold and Euclidean distances in the embedding space are virtually the same. This local relation enables the data analysis to use linear approximations, as Roweis and Saul (2000) observe. But is is also the reason why it is possible to create valid psychophysical functions *locally* with traditional methods, but to fail with a *global* analysis.

Nonlinearities that are sometimes puzzling from the perspective of Fechnerian psychophysics include cusps, the breakdown at extremes of Weber's law, negative masking and simple discontinuities.

Manifolds

An invariant manifold is the name given by mathematicians to a surface contained in the phase space of a dynamical system which has the property that orbits starting on the surface remain on the surface throughout the course of their dynamical evolution (Wiggins, 1988, p. 26). To visualise an example, consider a linen handkerchief which, when laid flat on the table, is a two-dimensional rectangle. Scrumple it up a bit and put it into a transparent plastic rectilinear box. To give the coordinates of any point on the handkerchief within the box needs a vector of three terms: the x, y, z coordinates defined with reference to the sides of the box. But distances running tightly on the surface of the handkerchief, and not jumping through the space in which it is contained are still distances in a two-dimensional topology.

In psychophysics, the nearest analogue to a manifold is a response surface; examples using Γ theory (Gregson, 1988, 1995, 1998) might also be seen to resemble a slightly scrumpled handkerchief. There are earlier examples, Osgood's (1953, p. 532) transfer and retroaction surface in serial and transfer learning theory is a manifold. But some parallels between mathematical definitions and the properties of real psychophysical systems can, with caution, be pushed a little further if we are prepared to accept that what happens in sensory and perceptual systems is governed by some nonlinear dynamical principles. In the psychophysical example, there is always some noise present, which means that data points that ought to lie in the response surface will lie near to it, like insects buzzing around a honey-soaked cloth. If and only if we know the equation that defines the response surface, then we can partition data into points on the surface and points clustered around the surface, just as in the simplest almost degenerate case of a single psychometric function (or item characteristic curve) we can linearise the plot and use regression theory to estimate goodness-of-fit.

The mathematical theory enables us to go further and find that the dynamics of points near to an invariant manifold M, which are attracted to or repelled from the manifold, are themselves in manifolds, called respectively the stable and unstable manifolds of M. Only if one understands all the dynamics of M and its associated stable and unstable manifolds can one describe what the system will do as it evolves in time. To parallel this in behavioural terms, one has to understand the capacity of a system for self-correction and for its breakdown under overload to understand it properly. Studying it only under stationarity, which is studying it on the invariant manifold, is not enough. Our interest in transients, then, is an interest in the local trajectories that run onto an invariant manifold or run away from it, but, preferably, onto it, as we are more concerned with bringing systems under control than watching them self-destruct.

Identification of local manifold regions

There are various ways in which the actual response surface of a system might be identified in the region in which we have data, when that response surface is treated as part of the invariant manifold of a nonlinear dynamics.

Reconstruction of local trajectories onto the surface, where in a small region linear mappings are admissible, has been developed by Roweis and

Saul (2000). Alternately, one may create a model of a global region and
then differentiate the surface to find areas of maximum instability, which
are areas with the maximal information of potential interest for process
control. Let us consider what the failure of Weber's law tells us, as dynamicists. Forget, for the moment, that it is a unidimensional description of
system sensitivity. It tells us that there exists a middle range within which
sensitivity is informative, and end ranges where responses change only
coarsely with stimulus changes. This is a section through a manifold which
has a gradient in one region and is almost flat at the edges. The manifold is
bounded because biological systems do not responsd to zero or to infinity,
they only operate in a narrow window of physical inputs. The manifold
can be locally differentiated; that is, at any point, its surface x can be replaced with another surface whose values are the maximum local gradi-
ent $\delta x|_x$ at that point. This analysis is outside time; it is global and about
the end points of all the trajectories that have terminated on the invariant
manifold. If the surface of the manifold is viewed from the viewpoints of
all three of the embedding dimensions then the derivative surface has at
any point x, y, z a vector $\delta x, \delta y, \delta z|_{x,y,z}$ and the steepest gradient is a vec-
torial resultant of these three gradients (Weatherburn, 1927).

A process arrives on the invariant manifold if it is governed by the
stable manifold. The stable manifold is exposed in the time series which
are generated in the attractor basin. If the attractor dynamics are chaotic
then the largest exponent in the vector of Lyapunov exponents is positive,
but as the dynamics do not explode some or all of the remaining exponents
must be negative; the process stretches on one dimension while it shrinks
on another, staying within a bounded space. For this reason, the largest
exponent, if positive, is taken heuristically as sufficient evidence of the
presence of chaotic dynamics in the system. The Lyapunov exponents are
directional, but they are not the same thing as the local gradients on the
invariant manifold; one set is global in time and the other is local in space,
so both are needed for a coherent system description.

What, then, is the link between the time series described in the introduction and the time series of the stable manifold? They are not derived
from quite the same sort of experiment; for example, if both EEGs and

perceptual responses are collected in the same experiment, the time scale of the EEG recordings is relatively fast: one data point per 3 msecs in the gap between successive perceptual responses made slowly at some seconds apart. Preserving the finer distinctions of our terminology, the EEG series is an event series (it has to be if the FFT approach is to be used to get an energy spectrum in the frequency domain), whereas if the perceptual responses are paced by the subject's own response latencies and are of detection ('yes-no') form, then the response series is a time-interval series.

One sort of model that has been explored in stochastic representations is an accumulator, where the cumulative fast covert activity ongoing during the EEG series reaches a threshold or an absorbing barrier and then releases the overt slow response (Heath, 1992; Vickers, 1979). It is quite possible for the dynamics at one level to be linear and to be nonlinear at the other (Wright & Liley, 1996). If the system is already on the invariant manifold, then any time series is an orbit on the manifold; it may or may not be represented by a random walk in a tiny neighbourhood, but cannot be random over a wider neighbourhood, because of the connectedness of the manifold. However, the manifold may incorporate catastrophes and so time series of orbits may have transients through singularities.

The fundamental problem with short records is that they may be generated in a diversity of ways:

- partly a stable manifold,

- partly later an invariant manifold,
  (the probability of beginning a data series on an orbit on the invariant manifold is asymptotic to zero if the process is perturbable and needs real time to come to stability.)

- or stochastic high-dimensionality perturbation of a linear process.

In any of such processes, there may or may not be stationarity in the dynamics[1]. The nonstationarity may be a form of jumping between the

---

[1] In the limit, the process is already on a point attractor, which is a manifold of zero di-

components already listed, or even jumping from one manifold to another. If the attractor is multi-lobed, as some of the classical cases in the physical literature are known to be, so that they jump intermittently from one lobe to another in the phase space, then even uncertain identification cannot result until one or more jumps between stable regions has occurred. Jumps between stable regions on a manifold are assumed to be catastrophes in various social psychological studies (Guastello, 1995) but jumps between basins of attraction in a delayed attractor's dynamics may simply be the results of jumps in an external forcing function (Gregson, 2001).

The idea that a time series could be generated by a set of states with conditional jumps between them was explored by Tong in the SETAR (Self-excitatory autoregressive) models. Another approach is to use stochastic differential equations, with jumps; under some conditions these converge to a Markov structure, a semigroup (Horbacz, 2004). In SETAR, any one-state evolution was modelled by an AR process of low (i.e. finite) order, but if the states were generators of orbits on a nonlinear manifold this would be conceptually inadequate. The difficulty is to model the transition rules between states or between corresponding stable (that is, stationary) regions, which is why the assumptions of the simpler catastrophes are so plausible and tractable.

Treating as Time Series

A little recapitulation will not come amiss, using a Markov representation and symbolic dynamics.

---

mensionality, and, from the psychologist's external viewpoint, we have complete response stereotypy.

$$
\mathbf{C} = \begin{pmatrix}
0 & 1 & 0 & 0 & \cdots & \cdots & 0 & 0 & 0 \\
0 & 0 & 1 & 0 & \cdots & \cdots & 0 & 0 & 0 \\
0 & 0 & 0 & 1 & \cdots & \cdots & 0 & 0 & 0 \\
0 & 0 & 0 & 0 & \cdots & \cdots & 0 & 0 & 0 \\
\cdots & \cdots & \cdots & \cdots & \cdots & \cdots & \cdots & \cdots & \cdots \\
\cdots & \cdots & \cdots & \cdots & \cdots & \cdots & \cdots & \cdots & \cdots \\
0 & 0 & 0 & 0 & \cdots & \cdots & 0 & 1 & 0 \\
0 & 0 & 0 & 0 & \cdots & \cdots & 0 & 0 & 1 \\
1 & 0 & 0 & 0 & \cdots & \cdots & 0 & 0 & 0
\end{pmatrix}.
$$

$\mathbf{C}$ is a Markov transition probability matrix containing one cycle through the states $1, 2, 3, 4, ..., n, 1, 2, 3, 4, ..., n, 1, ...$ It is the cycle that orders the states, not their intrinsic identities. Frobenius (1912) specifically discussed this form of non-negative square matrix. The process that $\mathbf{C}$ represents in symbolic dynamics is cyclic but not necessarily periodic. In order to decide if it is strictly periodic, and then compactly representable in the frequency domain by a Fourier expansion, it is necessary also to know the probability distribution of dwell times in each of the cells labelled with a '1'. If, for example, all of these distributions are exponential, of the same form, and decaying with a long tail, then it can be periodic as well as cyclic. Matrix $\mathbf{C}$ is not a probability transition matrix as it stands, but a state occupancy matrix, the distinction Parry is cited as drawing in Chapter 2. But, in the limit, it can be interpreted as a completely deterministic probability matrix. It is not then ergodic. Let us relax a little the form of $\mathbf{C}$ and see what happens. Let m be the rank of $\mathbf{C}$. Put $q = (1/m) * (1 - m)p_{i,i})$ and fill the matrix $m \times m$ with q where we had zeroes. The matrix is then ergodic, and maximum entropy.

$$
\mathbf{C_m} = \begin{pmatrix}
q & p_{1,1} & q & q & \cdots & \cdots & q & q & q \\
q & q & p_{2,2} & q & \cdots & \cdots & q & q & q \\
q & q & q & p_{3,3} & \cdots & \cdots & q & q & \\
q & q & q & q & \cdots & \cdots & q & q & q \\
\cdots & \cdots & \cdots & \cdots & \cdots & \cdots & \cdots & \cdots & \cdots \\
\cdots & \cdots & \cdots & \cdots & \cdots & \cdots & \cdots & \cdots & \cdots \\
q & q & q & q & \cdots & \cdots & q & p_{m-2,m-2} & q \\
q & q & q & q & \cdots & \cdots & q & q & p_{m-1,m-1} \\
p_{1,m} & q & q & q & \cdots & \cdots & q & q & q
\end{pmatrix}
$$

We can now derive the stationary state vector, when $C_m = 7 \times 7$, and the eigenvalues, from C_m, which are, for $q = (1 - p)/(m - 1)$ and if $p = .8, q = (1 - .2)/6 = .03\dot{3}$, as it is asymmetrical some eigenvalues are complex conjugate pairs, that is

E: 1.00, -0.694±0.334i, 0.480± 0.602i, -0.1713 ± 0.750i

The stationary state vector (SSV) is a rectangular distribution. If the cyclic dynamics are disrupted by a periodic transient, such as is done by putting the fourth row of C_m as .04, .04, .04, .04, .40, .40, .04, then the SSV is quite different, and after relatively slow convergence shows a peak for state 4.

Tremor Series

It could be argued that the comparison of different measures of dynamics will vary with stronger erratic characteristics of the time series used, and so for comparison a series with very different appearance, and irregular transient extreme deviations, unlike, for example, some normal cardiac series, is treated analogously to see what happens. We have chosen records of the motor control fluctuations and hence the failures that characterise Parkinson's disease, but without choosing here to model the physiological dynamics within limbs and brain that are the basis of what is externally recorded, simply to look at the time series themselves. The example chosen is s14r45of.d from the PhysioNet archives; only the first 4000 steps are used, in blocks of 400. The dynamics are on visual inspection nonlinear and non-stationary. Deviations can be peaks isolated in time, or come in sustained bursts.

Figures 3.1, 3.2 and 3.3

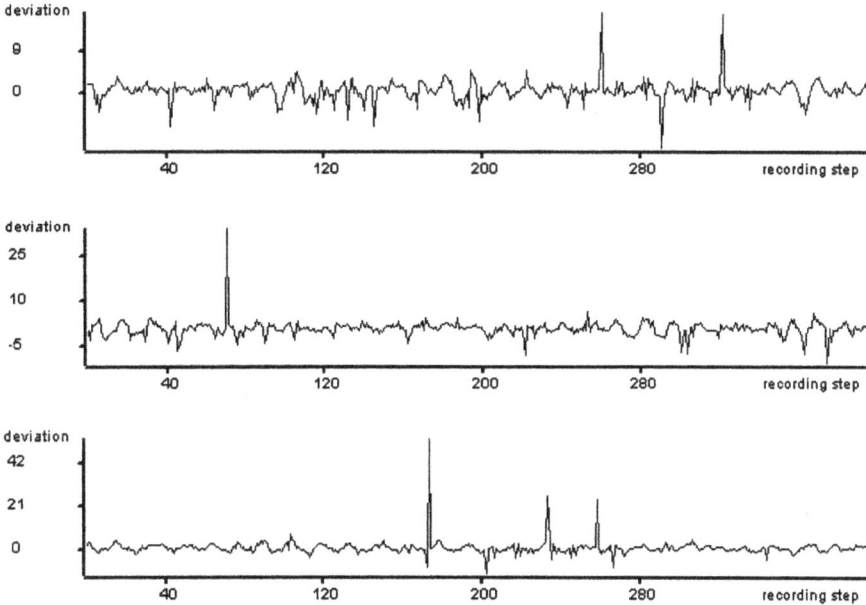

The raw data of Parkinsonian blocks 2, 4 and 10, illustrating transients. Vertical scales vary.

These series are wildly fluctuating, and their irregularity is reflected in most of the computed indices. The distribution of the dependent variable (as in the vertical axes in the graphs) is nothing like Gaussian, and would not he readily represented in the frequency domain, as Fourier analyses cannot handle local aperiodic spikes. Autocorrelations would be similarly misleading except as long-term averages. It is possible to treat this sort of series in a Markov chain filtering, as we did in the introductory example's Epochs. It is possible that the leaky bistable matrices introduced by

Mitchener and Nowak (2004) are a candidate model.

There appear to be two levels of tremors, low amplitude high frequency and intermittent aperiodic high amplitude, which coexist. High-pass and low-pass filtering could be used to partial these out. Parallels with the dynamics of sleep-wake cycles involving modelling by two coupled oscillators can lead to similar algebra (Strogatz, 1986).

Table 3.1: Descriptive statistics of Parkinsonian Tremor series

Block	LLE	mean	Kurt	ESf	D2	H	ApEn
1-400	+.034	.812	133.41	-.3897	1.835	.1647	1.112
401-800	+.078	.717	19.63	-.5572	1.806	.1928	1.122
801-1200	+.083	.529	6.43	-.5027	1.806	.1937	1.120
1201-1600	+.052	.725	71.66	-.5578	1.780	.2200	1.188
1601-2000	+.009	1.033	156.22	-.5025	1.932	.0684	1.092
2001-2400	+.073	.831	5.93	-.6258	1.788	.2118	1.245
2401-2800	+.096	.640	5.82	-.6642	1.790	.2103	1.250
2801-3200	+.065	.937	23.80	-.6354	1.780	.2201	1.259
3201-3600	+.080	.756	5.17	-.6146	1.836	.1641	1.219
3601-4000	+.023	1.106	122.78	-.4891	1.750	.2498	1.111
–	–	–	–	–	–	–	–
cftv	.449	.207	1.045	.145	.298	.250	.054
regr	1	-4.94	-1874.52	-1.89	-5.08	0.78	1.44
corr	1	.825	.910	.660	.289	.461	.635

Key: Block: beats in sequence. LLE: largest Lyapunov exponent. mean: average i.b.i. Kurt: kurtosis. ESf: entropic analogue of Schwarzian derivative. D2: fractal dimensionality. H: Hurst index. ApEn: approximate entropy ($k = 10$, $\delta = .02$). cftv: coefficient of variation ($\sigma/|\mu|$). regr: slope of regression of variable on LLE. corr: product-moment correlation of LLE with variable. For ESf, $k = 10, \delta = .005$

It now appears, from other quite separate data on oscillations in Parkinsonian tremors from a gyroscope-based recording instrument (Motus bioengineering), that the local LLE is related in its fluctuations to comparable fluctuations in the local ESf and these variations are compatible with differences in the ESf arising between random and edge-of-chaos activity. The bispectral approach also identifies local epochs with anomalous dynamics. Non-stationarity does not create an absolutely critical objection to using all these methods, rather the non-stationarity is intrinsically interesting in biological processes.

For a brief comparison, exploring some data from the Motus recorder, (kindly provided by Dr. Felsing) in four blocks of 200 steps each, a single entropy value from each $bESf$ matrix is derived; the original data series are left in their raw units, and the $b(n, m)$ values computed. The notation $b(n, m)$ for $B(n, m)$ is used to emphasise that these are short sample and not infinite summation estimations. The $b(n, m)$ are then rescaled to (0,1) over the whole triangular matrix, so that each $0 < b^*(n, m) < 1$, and the $\sum^{n,m} b \cdot log(b)$ value is used as a single index of the matrix information. Call this $Ent : bESf$.

The Motus series are dominated by a sinusoidal carrier signal and are hence quite stable over successive subsamples.

Markov and Eigenvalue representations

Each subset time series y can be converted to a 5-state symbolic series Y, and its associated Markov transition probability matrix then computed. This conversion to non-metric encoding is the shift from y to Y in Chapter 2. The choice of five states is arbitrary, but sufficient for our exploration of instabilities in the dynamics. The last column is an estimate of the Stationary State Vector for the matrix.

If the data in Block 2 are taken as generated by a basic symbolic process X unperturbed by transients, then the SSV in that block can be used as

Table M
Entropy values based on bESf and Esf
of Motus series

Sample	Ent:bESf	ESf
1	4.1465	-.1637
2	4.8012	-.0541
3	4.8221	-.0431
4	5.4598	-.0251

Matrices of maximum and minimum differences of deviates in the cells of the four bESf triangular matrices
Maxima (bolface) are written over their corresponding minima.

4.2737	**2.4720**	**.7191**	**-.0059**	**-1.5805**
3.9950	2.4511	.5646	-.1096	-1.9004
2.5889	**.7240**	**-.1613**	**-2.0440**	
2.5460	.5853	-.3210	-2.1011	
1.7894	**-1.314**	**-1.8984**		
1.6936	-1.4740	-1.9878		
-.2319	**-2.3957**			
-.3006	-2.4869			
-1.9232				
-2.1691				

---

estimates of $p_i(X|\theta)$, $i = 1, .., 5$ in the Kullback-Leibler expression

$$I(Y, X) = \sum_{i=1}^{5} p_i(Y) \cdot log\left(\frac{p_i(Y)}{X_i|\theta}\right) \qquad [3.1]$$

and that may be taken as a relative mismatch distance (not a true metric distance, because of asymmetry) between theory in Block 2 and data in

Tables 3.2: Markov matrices for blocks.

The transition probability matrix for Subset 2 in a 5-state encoding.

.205	.436	.256	.103	.000	: .0977
.130	.426	.374	.070	.000	: .2882
.054	.210	.613	.113	.011	: .4662
.105	.158	.316	.421	.000	: .1429
.000	.500	.500	.000	.000	: .0050

Eigenvalues: 1.000, .316± .015i, .021, .012

The transition probability matrix for Subset 3 in a 5-state encoding.

.074	.148	.333	.444	.000	: .0676
.071	.143	.393	.357	.036	: .0675
.106	.041	.463	.382	.008	: .3076
.047	.062	.209	.645	.038	: .5323
.000	.100	.200	.700	.000	: .0251

Eigenvalues: 1.000, .283, -.063, .052± .024i

The transition probability matrix for Subset 8 in a 5-state encoding.

.200	.200	.533	.067	.000	: .0375
.024	.293	.561	.122	.003	: .1028
.034	.086	.776	.100	.003	: .7268
.019	.019	.635	.327	.000	: .1303
.000	.000	1.000	.000	.000	: .0025

Eigenvalues: 1.000, .219± .046i, .165, -.006

The transition probability matrix for Subset 10 in a 5-state encoding.

.000	.714	.286	.000	.000	: .0178
.011	.849	.134	.004	.004	: .7053
.029	.340	.621	.010	.000	: .2694
.000	.500	.500	.000	.000	: .0051
1.000	.000	.000	.000	.000	: .0025

Eigenvalues: 1.000, .496, -.010± .051i, -.006

the other blocks. For block 3 compared with block 2, $I(Y, X) = .7064$, for 8 with 2, .2410, and for 10 with 2, .6167. Block 3 has the greatest deviance, so the lowest relative posterior Bayes probability of being generated from the dynamics in block 2 (Burnham and Anderson, 1998).

A possible further type of exploratory data analysis on series with suspected non-stationarity arises here and, for interest, is described. The intrinsic asymmetry of $I(Y, X)$ means that for a set of $m \in M$ subsets or sequential non-overlapping data blocks there are $M(M-1)$ comparisons, the auto-comparisons $I(m, m)$ are all zero by definition. This can be set out in a asymmetric square matrix, with leading diagonal (trace) of zeroes, whose eigenvalues will include a dominant term. Using for illustration the four matrix SSVs in Tables 3.2, we have then from [3.1] the $I(Y, X), Y, X, \in M$ matrix

Y→:	Bl. 2	Bl. 3	Bl. 8	Bl. 10
X:Bl. 2	0	.7064	.2410	.6276
X:Bl. 3	.6253	0	.6292	2.2592
X:Bl. 8	.2889	.7491	0	1.5309
X:Bl.10	.9284	3.6118	1.4041	0

with eigenvalues $E : 3.817, -2.965, -0.635, -0.217.$

Higher-order Dynamics

The $bESf$ matrices are computable for each subset and the triangular form shown in Chapter 2 can be rewritten in square symmetrical form, but the entries are not necessarily positive (Gregson, 2002). The leading diagonal cells are taken to be the mean of the off-diagonal cells. This convention has been checked to show that it does not materially effect the eigenvalues. The eigenvalues for each of the ten subset matrices are given in Table 3.3. As the matrices are symmetric by definition none of the eigenvalues are complex.

There now appears to be a transient perturbation around block 4. Due to local memory in the system the first-order transient in block 3 leaves

Table 3.3: Eigenvalues of the ten reflected bESf matrices

Block	E1	E2	E3	E4	E5	E6
1	-2.1975	-0.2343	0.2223	0.1822	-0.1509	-0.0064
2	-2.3814	0.2233	-0.2166	0.1412	-0.1041	-0.0287
3	-2.2254	0.2756	-0.2324	0.1694	-0.1307	0.0618
4	-1.9827	-0.3001	0.1713	0.1052	0.0489	-0.0077
5	-2.7091	-0.3209	0.3162	0.1739	-0.1143	-0.0499
6	-2.0382	-0.3044	0.2288	0.0521	0.0426	-0.0279
7	-2.4898	0.2460	-0.1660	-0.0904	0.0660	-0.0468
8	-2.6348	0.2136	-0.1600	-0.0534	0.0323	-0.0273
9	-2.4972	0.3059	0.2358	0.912	-0.8158	-0.0810
10	-2.4093	0.2931	-0.2849	0.1507	-0.1379	-0.0121

traces that show later. Such traces are sometimes called persistence. Yet another note of caution is in order: if one employs more conventional stochastic time series analysis on nonstationary data, it is possible to make identification errors, positive and negative, and only some models are properly sensitive to the detection of persistence (McCabe, Martin and Tremayne, 2005). It is only sensible to search for persistence if there exists some plausible reason to believe that the substrate physiology or memory processes have the capacity to support such persistence. In this sense, the psychology provides boundary conditions within which a statistical model has to be framed. Relevant models of working memory have been revised by introducing nonlinear dynamical networks in the relevant theory and there are thus open and unsettled questions about how persistence might be stored (Machens, Romo and Brody, 2005).

This interesting and unstable tremor series is revisited again in Chapter 10 for a further and slightly different analysis, which complements what is explored here.

Chapter 4

Inter- and Intra-level Dynamics of Models

In the period since the early 1980s, the amount of research published on the nonlinear dynamics of systems, including networks, has so expanded that it is impossible, even if one had the necessary erudition, to comprehend and synthesise effectively its implications for the psychologist. Admittedly, only a tiny fraction of this work is specifically addressed to problems which plausibly arise in considering brain and behaviour, but, even then, one has to selectively sift through results to discern what has implications for data analysis and theory construction.

Historically, one may trace an evolution and a shift in the style of modelling: from the simplest networks with two layers, through multi-layered nets, to network dynamics with stable and unstable basins of attraction (Killeen, 1989, Thelen and Smith, 1994), on to systems with two or more simultaneous time scales, supporting fast and slow oscillations (Cveticanin, 1996) and giving more consideration to what is becoming learnt about mid-brain information transmission, encoding and storage (Jensen and Lisman, 1996). On the one hand, the increase in the power of computation to the massively parallel has partly closed the gap between the effective degrees of freedom of the functioning brain and the simulation of at least one or more of its subsystems; sensation, perception, memory, decision and choice. On the other hand, the availability of newer non-intrusive brain scan technologies, PET and MRI to augment the limited insights provided by EEG, has cast irreversible doubts on simplistic mappings of consciousness to neurological activity (Revensuo and Kamppinen, 1994; Henson, 2005).

The processes which we can observe at the behavioural level are the consequences of mass action in the central nervous system; pathways with multiple serial and parallel cross-coupled interaction are recursively activated and can be not only nonlinear in their dynamics but also not stationary in the parameters of any model which captures, at least, the robust features of their qualitative dynamics. There are at least three distinguishable ways in which we may constrain the evolution of a trajectory. Modelling the single neuron is not what we are after. Modelling systems which can still function in much the same way when a fraction of their connectivity graph is deleted is of prime importance.

The use of attractor neural networks (ANNs, see Amit, 1989) as psychophysical metaphors, with capacity to classify, remember and learn, has, unsurprisingly, shown that simple nonlinear couplings over local arrays in space and in time can generate a diversity of input-output mappings and can be sufficient substrates of Gestalt phenomena (Gregson, 1995, Ruhnau, 1995). Some phenomena created are static, some are transient, some are locally predictable, and some are irredeemably stochastic. If we choose to start with the physicist's spin-glass theory and then progressively relax some of its structural assumptions, it is possible to approach a metaphor of some neurobiological mass action. Even that intrinsically over-simplified ANN exhibits the switching between stability and transitory eruptive phenomena, which can also emerge from a possibly infinite collection of alternative models which at their core involve nonlinear recursive mappings evolving through time; in short, they exemplify an aggregate of trajectories, not a single trajectory. This aggregate is sometimes autonomous and sometimes slaved by inputs from the environment. The distinction links into what has been called *contextual guidance* in multivariate neural networks (Kay and Phillips, 1996).

Again, we are faced with a multi-level problem, as in the title of this chapter, but now the upper or dominant level represents constraint by the organism's metabolism and not directly by the environment. Of course, in turn, the environment can modify the metabolism, via nutrition, but the time scale of that is vastly slower than the processes in milliseconds which concern us here.

I would like here to explore one aspect of trajectory generation that I had briefly mentioned earlier (Gregson, 1988) and never subsequently elaborated for its computational consequences. The basic idea is very simple, namely that if the dynamics, necessarily dissipative and irreversible, are the mode of functioning of a real system and not just a mathematical recreation, then what they represent can only be achieved, in real space and time, by consuming energy (and producing waste products, but let us leave that out of the story for now). For example, the brain needs glucose and the mechanisms of conversion of glucose to energy to keep going. We know that, in fact, the brain is a dominant user of energy in the economy of the total body. There are limits on how fast it can function, which have preoccupied psychophysicists since the mid-19th century, and it can run out of energy and become dysfunctional, or atrophy in some of its cells. There is always death on the eventual horizon.

One is therefore tempted to take a very simple starting point, not even a network in 3D but a single nested recursive channel, and see what happens to its quasi-stability, periodicities, and even its frequency spectral representation, if we embody in the model additional constraints on what are implicitly the changes which need energy to take place. These are the changes in the levels of the core system variables; these levels should be expressible as potential energy values. If one were to go initially in this argument to $n\Gamma$ cascades then the possibility of hyperchaos could arise and muddy the waters even more (Thomsen, Mosekilde and Sterman, 1991). The necessary algebra is derivable from the multichannel and cascade $n\Gamma^k$ precedents but is perhaps untidy.

However, without any of these entropy-bounding considerations we observe in a nonlinear system the diversity of dynamics which are representable by the regions and boundaries of Julia sets and by the homoclinic and heteroclinic orbits in phase space diagrams.

If the dynamics are within an environment whose series of inputs might push the process across a heteroclinic orbit (Wiggins, 1988, p. 182), then some additional constraints on the dynamics are needed if the organism is to survive. In mathematical terms, the introduction of these constraints on the components of a multiple channel system would open up the possibilities, already noted by Rössler (1991), of a hierarchy of types of chaos (Baier and Klein, 1991) some of which are not analytically tractable even though they can be readily simulated. They are also almost impossible to identify from outside without parametric knowledge of the constraints themselves. To keep things simple, we are not going to consider the possibility of a real biological system sitting permanently in a heteroclinic tangle; from an evolutionary perspective, that seems unrealistic.

Putting new bounds on the rate of evolution of trajectories, and hence protecting a system from running into explosion, at the expense of forcing its evolution into only some subregions of its natural dynamics, has consequences which can be counter-intuitive. There is no psychophysically good a priori reason to restrict any model to analytically tractable structures, because there is no a priori reason to assert that such processes at present exist in the real organism (what is analytically tractable changes slowly and extends over time with the emergence of new mathematical ideas). For a demonstration that some qualitative phenomena can arise in a bounded nonlinear system, as opposed to showing necessary and sufficient conditions for them to arise, it is unhelpful to move prematurely to analytic treatments if our intended focus is on generating a diversity of externally-observable dynamics without recourse to parameter proliferation in the model core. Instead of pushing up degrees of freedom to capture more and more transients and complexities, we can do the opposite and constrain the freedom of the dynamics locally by imposing bounds on the derivatives in time in such a way as to introduce yet more nonlinearities, even to create singularities in the evolution of trajectories; but what sort of singularities is not something our intuitions can reliably anticipate.

Given that we have input variables, internal parameters not coupled to inputs, and the system's state variables, imposing bounds to stop degeneration onto either explosion or the equifinality of death means that we must constrain either limits of variability or rates of change and their higher derivatives. After some heuristic exploration, an extension of a Gamma (Γ) cascade (Gregson, 1992,1995) within various bounds has been created to show what choices face the theoretician. The simplest Gamma trajectory, like many other recursions in a complex function, Y, with the reals on the bounded interval (0,1), will with small parameter changes exemplify point, periodic, quasi-periodic and strange attractors, and have quite different dynamical evolution in its real and imaginary parts. As I am using NPD cascades as the working example here, the limits can be put on a, the real input, on the real and imaginary parts of the system's variable Y(Re,Im), and on the cascade re-scaling parameter κ. We may also consider imposing limits on the cumulative local sum of Y, as that is a rough measure of the rate of energy consumption over a short recent time window. Psychophysiologists are thoroughly aware of the phenomena of dead intervals in neural transmission following signal transductions from input to output; neural systems that need 'to stop and take a breath' are ubiquitous.

Using NPD operator notation (Gregson, 1988,1992,1995), for two cascaded stages subscripted $_{(1)}$ and $_{(2)}$,

$$\mathcal{C}: \quad U \mapsto a_{t(1)} \mapsto \Gamma_{a,e;Y} \mapsto \kappa Y_{(1)} \mapsto a_{t(2)} \mapsto \Gamma_{a,e;Y} \mapsto Y_{(2)}(Re) \mapsto x_t ....$$
$$[4.1]$$

and this is extended indefinitely to M cascade steps as required. It can be argued, from considerations of the duration of the temporal present in consciousness (Ruhnau, 1995), that a cascade is the minimal duration dynamic structure that can underlie binding in neuropsychology[1] but we do not need this interesting conjecture for the present exploration. It treats the evolution of internal representation of an original input as recursively restabilised nonlinear trajectories, as a cascade of Γ components which

---

[1] There is a possible parallel here between a cascade with boundary conditions imposed and what Lansner (1986) called a *running associative net*, which is a succession of active templates initiated, reconstructed, stabilized and finally terminated, fuzzy and context dependent. The ideas are advanced quite independently, however.

may each lie in a different region of phase space, alternating with a linear rescaling to convert back from a Y output to a new a input, whilst preserving some continuity in the system variable but subject to the new boundary conditions. In [4.1] κ is a multiplier on $Y(\mathrm{Re})$ outputs from the preceding cascade to counteract the effects of scale compression; the reasons for this, and its effects, are discussed in Gregson (1995)

There is also a parallel here with what is called a Šil'nikov condition (Šil'nikov, 1965, Friedrich and Uhl, 1992), which has been reported in EEG activity; there, an attractor is left and reentered recursively within one trajectory. However that dynamical model requires three dimensions and structural assumptions that have not been shown to be necessary or realisable in the present context.

So, imposing bounds on rate of change dY/dj (j is a time base but not in fixed msec units) with recursive linear rescaling and re-entry to the trajectory (which is what I have called a cascade step) is expected to show itself in the creation of new patterns of *transitions* between types of phase space dynamics, but the *dynamics themselves* will be qualitatively analogous and the path to chaos via period doubling, or odd periodicities, may persist within the dynamically unbounded segments of the evolution. We conceptualise the effect of the boundary conditions as intermittent, operating only when the system tries to cross a heteroclinic orbit into instability, so that their operation partitions the time series of the trajectory's evolution into segments each of which may be dynamically coherent and locally therefore analytically tractable in the sense of Γ algebra. But the series as a whole could appear random if filtered through, for example, linear time series modelling. Also the critical values of the parameters a, e, η in Γ which are associated with shifts in the dynamics will be confounded with the effects of setting the bounds from the higher level. Expanding [4.1] to show where the bounds operate, each Γ step is of the form of a cubic recursion of an internal state complex variable Y, within the time interval of one trial or one cascade step, in discrete time units j, has been written and generated as

$$\Gamma: \quad Y_{(j+1)} = -a \cdot (Y_j - 1)(Y_j - ie)(Y_j + ie) \qquad i = \sqrt{-1} \qquad [4.2]$$

with the recursion given a starting value Y_0 and running $j = 1, ..., \eta$. The stimulus input U maps affinely onto a, and e is an internal parameter. Y is the state variable in continuous existence before and after the Γ trajectory, which is itself a transient destabilisation of the system. The terminal real value $Y_\eta(\text{Re})$ corresponds to an observable response, if the process terminates there. The corresponding imaginary part of the trajectory $Y(\text{Im})$ is always second-order with respect to the real part. Original initial gain $a_{(1)}$ is the only direct link with the system's environment.

The restabilisation step used here, from cascade J to $J + 1$, is

$$a_{J+1} = Y(Re)_J \times (5.49 - 2.01) + 2.01\kappa \qquad [4.3]$$

where the numerical values are used to keep the recursion within bounds; they are arbitrary, within limits.

The structure of [4.1, 4.2, 4.3] creates two levels of recursion, within η steps of a Γ trajectory and within a cascade step J. That is, Γ is nested within the cascade recursion. Taking this structure one level deeper, a run of M ($J = 1, ..., M$) cascade steps can be one recursion within a higher-order loop K which, each time it is entered, reinitialises on $a_{(1)}$ by contacting the stimulus environment U again, but keeps the running values of $Y(\text{Re,Im})$ from the last M of K. Obviously U can change in time quite independently of the looped cascades of Γ trajectories.

Using an operator notation, we have to get from a stimulus U to a response Y_{obs}

$$Y_{obs} = K(M(\mathcal{C}))a|U \qquad [4.4]$$

and this evolves in $\eta \times M \times K$ time units. We now need to look at some of the time series properties of [4.4] when it is constrained by bounds on its energy consumption.

Though this sort of model was derived from considerations of external psychophysics and to a much lesser extent neurophysiology, it can in mathematical terms be regarded as a special case of the Cellular Neural Networks (CNNs), which are strictly applied mathematics and usually simulated by electronic circuits, even though their analogies with brain processes are optimistically asserted by their creators. By treating these analogies tolerantly, the wide range of mathematical results now available can be drawn on to provide insights into how the $n\Gamma$ cascades and nestings also may furnish a basis for learning, memory, and pattern recognition, discrimination and storage. (Brucoli, Carnimeo and Grassi, 1995, Thiran, Crounse, Chua and Hasler, 1995).

Intermittencies

It has long been known that EEG time series can exhibit sudden peak fluctuations, whose inter-peak interval frequency distribution may be stochastic; these, for example, occur in the diagnosis of epilepsy. It has also been noted in nonlinear dynamics that the evolution of attractors expressed as a time series can shown long runs of stable periodicity with sudden apparently interpolated episodes of aberrant fluctuations. These interpolations may themselves have a recurrent easily identifiable form which I have previously called *arpeggios*; examples are noted by Gregson and Harvey (1992). It has also been known for some time that two statistical time series can run apparently identically for a subseries of trials and then abruptly diverge; their differences are not identifiable only from finite sample realisations unless and until they exhibit local divergence. This is not necessarily an example of chaotic dynamics; it can occur with linear models.

Intermittent recurrent subsequences arising within relatively stable series are thus dynamically ambiguous; the standard methods of identification, based on ARIMA, FFT, or adaptive filters will either treat the subsequences as noise because they are aperiodic in a quasi-periodic environment, or set up a compound mechanism in which the intermittencies are generated autonomously from the background series and then summed with its spectra by, probably, a weighted linear rule. The distinction between tracking such a series locally with adaptive filtering for its control and predicting its long term evolution has to be kept in mind. The approach here is focussed on the generation of series, not on their identification when their underlying dynamics are unknown. The object is to see what qualitative forms emerge as a consequence of the boundary conditions, which could be misidentified by standard methods of data analysis.

The following examples are all with bounds imposed outside the Γ recursion, by the outer loops and *not* in the interval $j = 1, ..., \eta$. It follows that the cascades are not fully protected against explosion if the parameters a, e, η, Y_0 allow instability within the local trajectory of one cascade step J. Where parameters are not given, they are the same as in the previous example.

1: Consider [4.1 to 4.4] with the parameters and constraints

$e = .25, \eta = 10, K = 10, M = 10 \; \Delta^1 Y = 0.7, 10^{-5}$

$a_{(1)} = 5.051$

$\max \sum Y(Re) = 6$

$\kappa = 1.0$

After $K = 2, M = 10$ it runs for all $M = 10$ onto an attractor which is $a = 4.55..., Y(Re) = .73...$ that is nearly stable and very gradually increasing, $Y(Im)$ within a loop is dynamically constant, and $Y(Re)$ for $M = 1$ decreases with K. Within $j = 1, .., \eta$, $Y(Re)$ and hence a are both period 2.

However, we have only to alter η to 9 instead of 10, and because the Γ dynamic is past its first bifurcation, the attractor at $M = 10$ becomes as soon as $K = 2$ the point $a = 4.55427$ and $Y(Re) = .731113$. Instead the $Y(Im)$ series is now variable and aperiodic within a loop, and for $M = 1$ $a = 4.98650$ and fixed.

2: For comparison,

$e = .35, \eta = 10, K = 10, M = 10$

$a_{(1)} = 4.2$

induces an alternation at $M = 10$ of $a = 4.46, 4.85$ and $Y(Re) = 0.815, 0.704$ so that a periodicity of the recursion in K is observed, and $Y(Im)$ is aperiodic.

Now we can remove the outer loop computationally simply by putting $M = 100, K = 1$ and we obtain period 2 alternation in J of $a = 4.851, 4.456$ and correspondingly in $Y(\text{Re})$.

3: If the Γ parameters are set deliberately at the edge of a stable basin, almost on a heteroclinic orbit, then we get intermittency with an odd value coming up at apparently unpredictable intervals. For example, in

$$e = .35, \eta = 10, K = 1, M = 260$$

$$a_{(1)} = 5.34035$$

$$\kappa = 0.95$$

the a series fluctuates irregularly (aperiodically) between about 5.16 and 4.07, with odd values $\simeq 3.92$ observed at $J = 16, 48, 65, 76, 86, 113, 119, 128, 141, 149, 158, 167, 191, 197, 212, 234$, and 258 and the $Y(\text{Re})$ series has a corresponding low value at each of these points. This is a typical edge-of-chaos phenomenon; it requires no extra component in the model but only a careful choice of parameter values.

4: Moving well away from chaos in the reals, with

$$e = .25, \eta = 10, K = 1, M = 260$$

$$a_{(1)} = 3.4$$

$$\kappa = 0.95$$

we observe a strictly periodic intermittent episode, in a steady a and $Y(\text{Re})$ background. For nearly all values, after initial stabilisation, $Y(\text{Re}) = .71845$, but at $J = 41, 42, 43, 44$ we have $Y(\text{Re}) = .56, .64, .70, .72$. This little arpeggio episode repeats itself at $40J$ intervals.

Clearly, a diversity of qualitative dynamics becomes readily possible as soon as we admit nesting of recursions hierarchically and the core trajectory is nonlinear. The system in its repertoire of time series now includes

Figure 4.1: a values for looped cascade
$\eta = 10$. Recursion steps 0,....,260

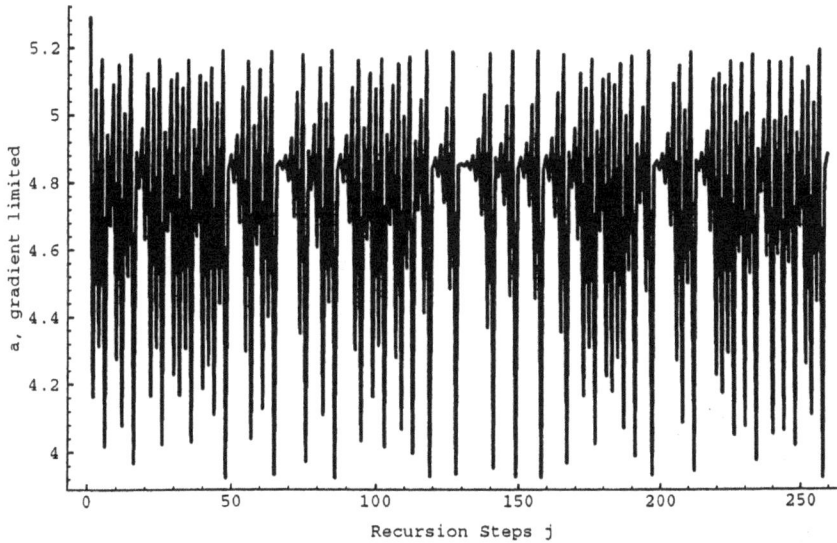

not just the steady state, periodic and path-to-chaos trajectories which are generated by [4.1], but the intermittencies of both predictable and unpredictable occurrence. The system is always deterministic in its generation, but may appear random in its outputs.

From the Gregson and Harvey (1992) results and other reported findings about nonlinear trajectories, we know that series with aperiodicity and local arpeggio sequences of identical recurrent form, but stochastic inter-event intervals can be generated by a number of well-studied attractors. The additional result here is the new type of series, with the baseline fixed and the arpeggio at fixed intervals, which are not the cycle times η, M, K but can be a multiple of them. This sort of series is a potential base for a biological clock, created without any special dedicated dynamics for such a clock in the system; it arises just because of the parameter settings in a narrow window. It is left open precisely how these control parameters can be brought into play and have their values set. Insofar as

Figure 4.2: a **values for looped cascade**
$\eta = 10$. Recursion steps 0,...,65

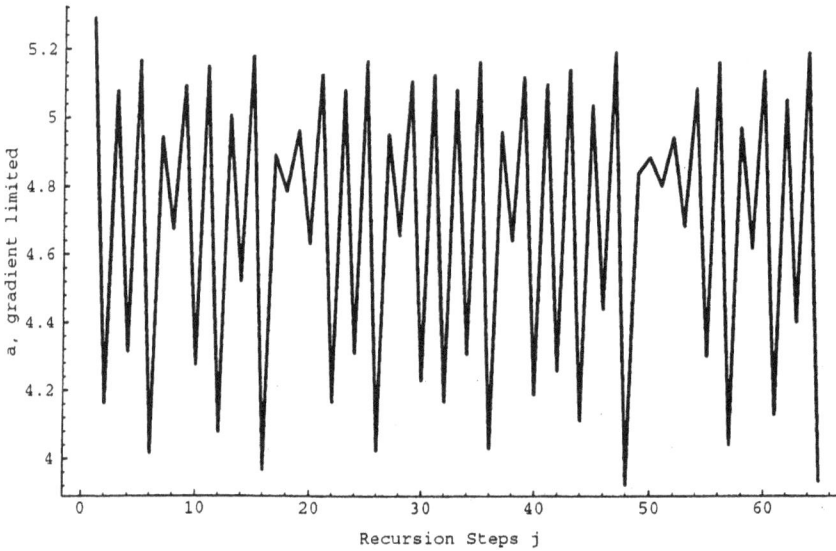

it is a problem of modulating shifts into and out of chaos, where chaos is on the edge of instability, the problem has received attention with a different approach concerned with the connectivity of the network in which the nestings are embedded (Doyon, Cessac, Quoy and Samuelides, 1993). Here we are assuming stable and completely connectively in the simulations; the assumption globally is not apparently necessary. An implication is that any recursive loop in the Γ family can, with constraints locally imposed, serve as a biological clock for a whole system of connected recursions. There is no need to have a specific locality for the time keeping process, and it can itself move about in slow time.

Figure 4.3: a values for looped cascade
$\eta = 10$. Recursion steps 0,....,260

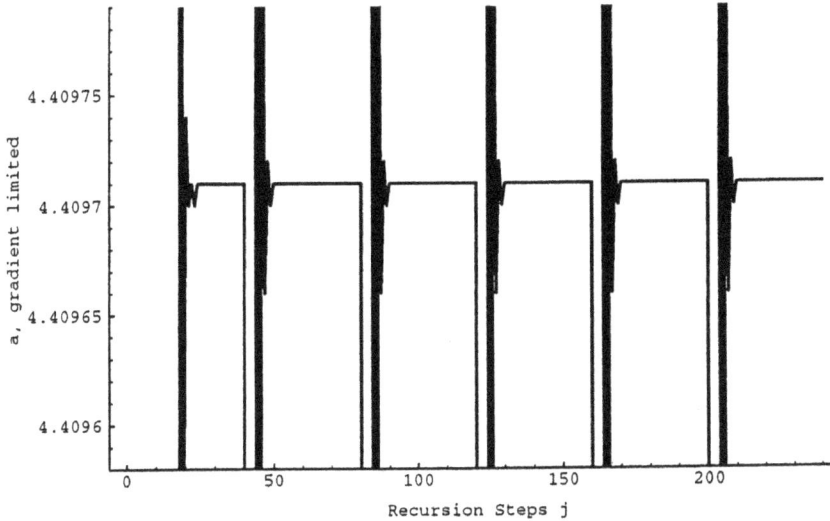

Synchrony and Binding

The approach here has been, so far, to start from nonlinear psychophysics and extend its structures to exemplify new recursive dynamics. The motivation has been partly to see where biological clocks might come from and partly to compare results in a wider sense with both cognition and with physiology. The second comparison turns out to be the relatively easier because of a detailed critical review by Singer (1993) which summarises the dynamics of neurological brain networks where they exhibit oscillations and synchronies. (The points raised by Singer are given in Appendix 1.)

If the oscillatory dynamics of two non-contiguous regions come into synchrony in some frequency range, then they are said to be *bound*. It is the so-called binding problem and the suggestion that binding is a necessary and sufficient condition for the emergence of consciousness that

Figure 4.4: a **values for looped cascade**
$\eta = 10$. Rescaled on a, Recursion steps 0,...,65

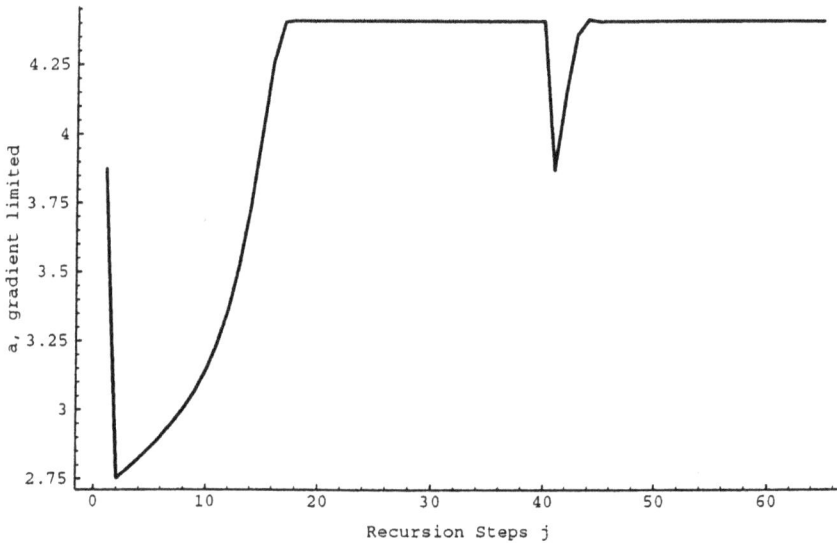

makes this property of nonlinear networks important for the psychologist. The formal mathematics of networks that oscillate exhibit synchrony, can bind, and may be a putative basis for consciousness were first sketched out in the 1980s; that work seems to be still unknown to most cognitive psychologists (Caianiello, 1987,1989, Palm and Aertsen, 1986). The need to attempt some three-way mutual compatibility between neurophysiology, conscious cognition, and mathematical representations of both, has been remarked on by various writers, but it does no harm to reiterate it here. The present innovation is to attempt to link a model that starts from psychophysics (and not from spin-glass nets or related physicalisms) to conditions for binding, and hence for the emergence of consciousness if the binding hypothesis holds.

There is, in the neurophysiology, a critical relation between delays in reciprocally linked groups and the establishment of synchronization. It is required to have a rapid succession of short bursts and an upper limit on the frequency of reverberation. The synchronous discharge creates strong simultaneous inhibition, and oscillations are a consequence of synchrony. Synchrony at a millisecond time scale is significant; it requires a temporal structure that allows the synchronous/asynchronous states distinction to exist with high temporal resolution and the establishment of synchrony over large distances. Singer (1993, p. 366) observes that both of these requirements can be met by paced oscillatory discharge patterns in a bounded frequency range. The properties match the dynamics of what has been observed in $(n \times n)\Gamma^k$ cascaded lattices (Gregson, 1995) if they are also nested in the bounded recursions of [4.1, 4.2, 4.4] set out here. To explore that situation more deeply we need to create an $n\Gamma$ extension of [4.1, 4.2, 4.4].

Bounded Cascades in 6-d partitioned NPD

These notes are not a simulation of a specific experimental configuration, but an exploration of what can arise in the dynamics of 6Γ **Case 2** with some parameter constraints. It is already known that multiple (one-to-many) solutions in the input-output mappings can arise, apart from those due to hysteresis. This example is now put into the nests of bounded recursions; it was previously used in an unpublished model of a problem in visual perception as a single 6Γ stage with some imbalances. As it is somewhat intricate in its dynamics it can serve as an introduction to systems where attractors co-exist. The qualitative features of output nearequivalences can be shown: In a 6-channel cross-coupled Γ system there are six gain parameters $a_1, a_2, ..., a_6$, and potentially six $e_1, ..., e_6$ internal parameters. These latter are reduced by constraints on cross-coupling, to $a_1, ..., a_6, \lambda_1, .., \lambda_3$.

The following table summarises the settings used here; the channels are in fact treated as being coupled locally in pairs.

This configuration means that channels 1 through 4 are fixed in their input settings, but channels 5 and 6 are variable and could float over a wide arbitrary range. For the current demonstration we do not need to invoke this plasticity; computations with all initial parameters fixed, chosen to be near to chaotic activity, are sufficient (compare Freeman, 1994). The purpose of this exploration is to look at the effects on outputs $\{Y_1, ..., Y_6\}$(Re) of the variations in a_5. Because of the cross-couplings the values of Y_1, Y_2 are functions of the total parameter set, and not just of a_1, a_2, e_1, e_2.

It can be shown that the behaviour of $|Y_2 - Y_5|$(Re) is under some settings of $\{\lambda\}$ multimodal. This is sufficient for the present heuristics. See Appendix 2 for computational details.

Table 4.1: Parameters having degrees of freedom
in the original program modified for this study

Channel	a	λ	a:fix or var
1	a_1	λ_1	fixed
2	a_2	λ_1	fixed
3	a_3	λ_2	fixed
4	$a_3 + .2$	λ_2	fixed
5	a_5	λ_3	variable
6	$a_5 + .2$	λ_3	variable

The cross-coupling between the six dimensions is only partial; they are grouped here into three subsets, by the operation of the λ coefficients as in [4.5], to define the (complex) e for each Γ as in [4.1].

$$
\begin{aligned}
e_1 &= (0, \lambda_1/\{\max(a_3, a_4, a_5, a_6)\}) \\
e_2 &= (0, \lambda_1/\{\max(a_3, a_4, a_5, a_6)\}) \\
e_3 &= (0, \lambda_2/\{\max(a_1, a_2, a_5, a_6)\}) \\
e_4 &= (0, \lambda_2/\{\max(a_1, a_2, a_5, a_6)\}) \\
e_5 &= (0, \lambda_3/\{\max(a_1, a_2, a_3, a_4)\}) \\
e_6 &= (0, \lambda_3/\{\max(a_1, a_2, a_3, a_4)\}) \qquad [4.5]
\end{aligned}
$$

A possible consequence of [4.5] is that the system as a whole can run into what Rössler (1991) has called *hyperchaos*. For the present purposes, the interest is that the system can run onto a series of point attractors; that is, a vector $\{a\}$ or individual Y can show transient perturbations independently of the concurrent dynamics of the other channels. These transients are like those identified in the example **3** for the simpler case already described. The rate at which the system as a whole, expressed in vectors $\{a\}$ or $\{Y(Re)\}$, goes to stability depends on the values of the loop constants in [4.4], which are common for the whole set of six dimensions, and on the e in [4.5] for the individual dimensions. Terminal values of a (and Y) can be unequal when initial values are equal and vice versa.

We must now address the problem of what happens if we use linear time series analyses methods in an attempt to identify periodicities, via autocorrelation spectra, in the component detailed Y series with $\eta \times M \times K$ terms. As might be expected, until the process has run onto an attractor or a set of attractors it is not stationary and estimations are confounded. Its hidden generators are very nonlinear, in [4.1] and [4.2], and are not recoverable from outputs Y or the restricted series Y_η with $M \times K$ terms. The fact that linear time series methods are dangerous in biological analyses is now well documented (Cutler and Kaplan, 1996). Another problem arises: the autoregressions are completely invalidated if we include the extreme realizations within the transients when they arise, and computational degeneracies result. An example is shown in Table 4.2, for recursions Nos. 192-202 with $\eta = 10, M = 5, K = 10, \kappa = 1.46, \{a\} = \{3.15, 4.70, 4.10, 4, 80, 4, 20, 4.50\}$, and $\{\lambda\} = \{1.00, 1.25, 1, 29\}$. There is nothing necessarily surprising in such phenomena; complicated multichannel processes can double back in their dynamics when their bases are nonlinear (Dawson, Grebogi, Yorke, Kan and Kocak, 1992).

Table 4.3 gives an example where the processes shown no explosive transients, but the six components vary considerably in their quasiperiodicity. If there were binding, then we would expect two or more series to be in phase and exhibiting a comparable spectrum of the autocorrelations. To check this we may set up the (triangular) matrix of cross correlations of the six vectors of 20 coefficients. It is also necessary, there-

Table 4.2:
Example of a local instability in one dimension, for $Y6$(Re)

192	0.655201	0.654684	0.673780	0.638777	0.518174	0.438628
193	0.822628	0.823239	0.810192	0.829268	0.827969	0.898943
194	0.653720	0.652882	0.665020	0.637210	0.670651	0.556254
195	0.822635	0.823236	0.812151	0.829139	0.878040	1.03977
196	0.653705	0.652890	0.661164	0.637503	0.529913	-0.287737
197	0.822635	0.823236	0.812769	0.829165	0.838031	1.16740
198	0.653705	0.652889	0.659937	0.637444	0.645619	-1.51008
199	0.822635	0.823236	0.812934	0.829160	0.883736	37.2423
200	0.653705	0.652890	0.659608	0.637456	0.511263	-318667.
201	0.513489	0.766225	0.705921	0.830226	0.684813	0.725012
202	0.470594	0.692762	0.686127	0.619342	0.716427	0.739834

fore, to examine the pair-wise cross-correlations of the raw series of 500 realizations. Cross-correlations are known to be informative in connected neural nets, but there are dangers in inference from externally observable properties by using what are in effect linear filters; "In the case of highly interconnected neural systems, for example when oscillatory behaviour is present, the response is far from linear. In such cases, therefore, any conclusions for the functional connectivity between two neural systems that are based solely on the presence of correlated activity are questionable." (Kalitzin, van Dijk, Spekreijse and van Leeuwen, 1997, pp. 73-74). Again, Liao and Sethares (1995, p. 14) noted that "In many applications where data are generated by a nonlinear mechanism, linear models are unacceptable and identification schemes fail This is mostly due to the fact that for general nonlinear systems there are no universally applicable models."

In Table 4.4, which is given as examples for the degrees of lag 0 to 4, (there is a series for each of the lags 1-20 computed), the leading diagonal cells would be filled with the autocorrelations of the individual MKT series. To obtain the roots of the total cascaded 6Γ when it is treated as one undecomposable series externally observed, we create the determinants of each of the lag matrices as in Table 4.4, with the autocorrelations added in, and use these determinants as the coefficients of a delay function which can itself be factorised. The method is given with examples in Gregson (1983). The factorisation of this process is to be compared with that of one or more component Γ trajectories.

A second approach is to impose a cut-off on the Y terms and treat those values above a critical level as the realisations of a point process. The interest is then in the statistics of the frequency distribution of inter-point intervals or of run lengths between crossing points. These do not necessarily have a periodicity which is a multiple or submultiple of the M, K, η terms, so that the periodicities of the point process are not a sufficient basis for system identification; this is relevant when considering EEG analyses of brain dynamics. We return to this aspect later.

The example deliberately created by the parameter settings for Tables 4.3 and 4.4 is intrinsically unstable and if it were not for the two outer loops would eventually explode. It is trivially simple to create very stable examples where the core Γ recursions run quickly in the Y reals onto point attractors. However, it is necessary to examine the roots of the factorisation of the polynomial whose coefficients are the sequence of the determinants of the completed matrices in Table 4.4. In Table 4.5 we give two cases, simply for illustration: the unstable one just created, and a very stable one with lower λ and a coefficents[2]. The roots for the imaginary components of the Y evolution have been shown as well, to emphasise the point that the dynamics are not exactly the same for the two parts of the complex system variable Y and, in general, the imaginary trajectories can be more turbulent, though second-order in magnitude, and are not postulated to be externally observable. Possibly, the numerical method has just failed to find a repeated conjugate root pair for the Y(Im) series in the first example. It would, however, be a mistake to equate the imaginary component with stochastic noise.

---

[2] The roots were obtained by using the Mathematica command NSolve.

Table 4.3: Autocorrelations in a externally stable case
$$\eta = 10, M = 5, K = 10, \kappa = 1$$
$$\{\lambda\} = \{1.00, 1.25, 1.29\}$$
initial a settings: $\{3.5, 4.7, 4.1, 4.8, 4.2, 4.5\}$ Autocorrelations lags 0 to 20;
read across lines, as text, in order

$Y(\text{Re})$ 1:
1.0000	0.5418	0.0840	-0.0095	-0.0122
-0.0168	-0.0142	-0.0201	-0.0129	-0.0196
-0.0074	0.2377	0.4253	0.2946	0.1289
0.0115	-0.0025	0.0041	-0.0026	0.0010

$Y(\text{Re})$ 2:
1.0000	0.5878	0.2806	0.0036	-0.0161
-0.0181	-0.0173	-0.0282	-0.0107	-0.0265
0.0085	0.0219	0.0124	0.0162	-0.0007
0.0108	0.0010	0.0073	0.0008	0.0036

$Y(\text{Re})$ 3:
1.0000	0.0373	0.5449	-0.1061	0.4652
-0.1397	0.3915	-0.1788	0.3218	-0.2277
0.2482	-0.2585	0.1822	-0.2981	0.1738
-0.3108	0.1623	-0.3190	0.1474	-0.3211

$Y(\text{Re})$ 4:
1.0000	-0.5416	0.7286	-0.5579	0.5897
-0.4539	0.4460	-0.3589	0.3263	-0.3085
0.2313	-0.3035	0.2605	-0.3207	0.2808
-0.3470	0.3008	-0.3629	0.3158	-0.3655

$Y(\text{Re})$ 5:
1.0000	-0.4581	0.6768	-0.4958	0.6225
-0.4907	0.5683	-0.4849	0.5155	-0.4796
0.4738	-0.4741	0.4338	-0.4774	0.4307
-0.4753	0.4284	-0.4740	0.4270	-0.4750

$Y(\text{Re})$ 6:
1.0000	-0.6879	0.7100	-0.6867	0.6885
-0.6677	0.6652	-0.6464	0.6410	-0.6223
0.6519	-0.6275	0.6506	-0.6338	0.6549
-0.6400	0.6608	-0.6470	0.6679	-0.6565

Table 4.4: Cross correlations 6×6 of lags 0 to 4 from Table 4.3
based on a run of 250 points
above diagonal, n to k, below diagonal reversed k to n

	n to 2	n to 3	n to 4	n to 5	n to 6
lag 0					
-.—	0.453	0.670	0.313	0.480	0.258
0.453	-.—	0.615	0.545	0.521	0.529
0.670	0.313	-.—	0.539	0.946	0.791
0.480	0.258	0.615	-.—	0.539	0.670
0.545	0.521	0.529	0.539	-.—	0.922
0.946	0.791	0.539	0.670	0.922	-.—
lag 1					
-.—	0.353	0.616	0.258	0.431	0.197
0.314	-.—	0.470	0.448	0.387	0.344
0.412	0.234	-.—	0.016	-0.038	-0.173
0.251	0.084	0.159	-.—	-0.030	-0.199
0.198	0.089	0.064	0.092	-.—	-0.412
-0.102	-0.232	-0.104	-0.219	-0.421	-.—
lag 2					
-.—	0.263	0.556	0.174	0.395	0.178
0.188	-.—	0.278	0.162	0.248	0.230
0.323	0.175	-.—	0.106	0.431	0.313
0.204	0.052	-0.070	-.—	0.284	0.391
0.059	-0.066	-0.029	0.231	-.—	0.472
0.431	0.332	0.149	0.264	0.494	-.—
lag 3					
-.—	0.158	0.480	0.121	0.329	0.113
0.110	-.—	0.072	0.055	0.044	-0.004
0.257	0.212	-.—	-0.232	-0.196	-0.328
0.160	0.068	-0.044	-.—	-0.189	-0.377
-0.019	0.001	0.033	-0.053	-.—	-0.466
-0.190	-0.258	-0.262	-0.299	-0.433	-.—
lag 4					
-.—	0.074	0.404	0.042	0.280	0.089
0.098	-.—	0.046	-0.073	0.034	-0.002
0.241	0.200	-.—	0.139	0.390	0.310
0.148	0.052	-0.053	-.—	0.327	0.402
0.042	-0.058	-0.035	0.262	-.—	0.455
0.386	0.325	0.206	0.304	0.475	-.—

The numerical values of the coefficients have been rounded off and should not be given precise meanings; what is important is the pattern of the roots estimated. In both examples, we have one real and two conjugate pairs from the few lags used in Table 4.4. The correlations in Table 4.4 decrease monotonically in many cells having the same location, as we run down the set of five 6×6 matrices. This implies, if we accept a causal interpretation, that there is weak binding between the channels in their quasi-periodic dynamical evolution. The drifting apart of trajectories in terms of their lag correlations might be related to the Lyapunov coefficients of the process. In both examples, we expect at least one complex conjugate pair of roots from the original Γ cores (compare [4.1]), and the rescaling of [4.3,4.4] creates another weak periodicity, of slower frequency determined by M, which is the periodicity forcing the 6Γ core back into its attractor basins. The raw time series to generate Table 4.3, for example, is computationally just a string of numbers with no given information about M, K, η. In this respect, it is like an ignorant external observer with only an output record.

We can retrieve the evidence of internal oscillations, but not necessarily an indication of precisely how they are generated. Here we have two temporally alternating mechanisms: one linear in the reals ([4.3]) and the other complex nonlinear ([4.2]). The two cases show complex conjugate pairs in both the stable and unstable half-fields together, a pattern which is sometimes associated with the presence of chaotic dynamics, although the example of Tables 4.3 and 4.4 appears basically homogeneous. To obtain a fuller picture of the system as it floats through its parameter space (there are at least nine parameters which can in principle affect the root pattern, though it is always the same system in its structure) requires the exploration of response surfaces and sections through them.

Comparison Control Conditions for the Dynamics

There are two simpler configurations which should be compared with the deliberately intricate examples summarised in Table 4.5. One is where all

Table 4.5: Determinants and roots for two cases,
one unstable and the other stable

The unstable case as in Tables 4.3 and 4.4

$\{\lambda\}$: $\{1.00, 1.25, 1.29\}$, initial $\{a\}$:$\{3.5, 4.7, 4.1, 4.8, 4.2, 4.5\}$

Y(Re), lags: 0	1	2	3	4
determinants: -309.7	-3.067	-.00625	$.583 \times 10^{-5}$	$-.279 \times 10^{-5}$

factored to give:

Y(Re) roots: real:	-1.1652	conjugate:	$-.446 \pm .980i$	$.529 \pm .312i$
Y(Im) roots: real:	-0.9987	conjugate:	$-.478 \pm .690i$	$-.477 \pm .692i$

A relatively stable case for comparison

$\{\lambda\}$: $\{0.90, 1.00, 1.10\}$, initial $\{a\}$:$\{3.2, 4.3, 4.1, 4.0, 3.5, 3.8\}$

Y(Re), lags: 0	1	2	3	4
determinants: -.00181	-.00317	.0406	.00175	.00228

factored to give:

Y(Re) roots: real:	-.0557	conjugate:	$-.892 \pm .394i$	$.419 \pm .904i$
Y(Im) roots: real:	-1.0238	conjugate:	$-.264 \pm .845i$	$.458 \pm .486i$

---

the channels are identical with uniform cross-coupling, and the other is where there is no cross-coupling, the **Case 1** of Gregson (1992,1995), so that e values are set in [4.1] independently for each channel.

The dynamics associated with the 6Γ **Case 1** example are apparently quite different from those for the corresponding cross-coupled **Case 2** example, which has otherwise the same parameters. The single conjugate paired root exists as in [4.2], but the remaining roots are all real and of mixed sign, indicating mixed stability. The implication is that there are structural changes in the internal dynamics of an $n\Gamma$ system which can be externally identifiable.

This is not the only distinction that emerges readily between the two structures in $n\Gamma$, **Case 1** with parallel channels having fixed e, and **Case 2** with cross-coupling via the $\{\lambda\}$ which in effect alter their operative e values with collateral inputs.

It is computationally trivial to show that setting arbitrary cut-offs levels on $Y(\mathrm{Re})$ for each channel will create quite different inter-crossing interval frequency distributions for the two **Cases** and that the point process series differ markedly from one channel to another within one case. Further, distributions that are regular but not multiple of M, K, η can be observed. Some data from one example are given in Table 4.8:

Serial Hypercycling

So far the cascading had been within each Γ process and the six processes are in parallel evolution, their trajectories are cross-coupled through [5] only. An alternative structure is where each trajectory cascades onto another dimension's Γ and the closed loop at level K is recursively through an ordered subset of the available dimensions. For example, the cascade beginning in [4.6] is serially and recursively through dimensions 1,2,3,4,5,6,1,2,3,...

$$\mathcal{C}: \quad U \mapsto a_{t(1)} \mapsto \Gamma_{a_1,\lambda_1;Y} \mapsto \kappa Y_{(1)} \mapsto a_{t(2)} \mapsto \Gamma_{a_2,\lambda_2;Y} \mapsto Y_{(2)}(Re)$$

$$\mapsto \ldots\ldots x_t\ldots\ldots \qquad\qquad [4.6]$$

and we may call this path a *hypercycle*. But there are N_{HC} possible such hypercycles, where

$$N_{HC} = \sum_{h=1}^{6} h! \binom{6}{h} \qquad\qquad [4.7]$$

and these are not externally distinguishable, except in their recursive periodicity, which is $h(\eta + z)$ in time units if we arbitrarily take the restabilisation process in [4.3] to have $z\,j$ units, when the six component Γ recursions each run onto a stable attractor.

But if one or more does not, because it has gone into chaotic dynamics, then the hypercycles of order h are not necessarily commutative. It is only if they are not commutative and we have full information about starting values $\{a\}$ that identification might be possible from the latencies and the terminal values.

The extension to 2D and 3D Lattices

Obviously psychologists are interested, at the very least, in having accounts of 2D processes in vision and 3D processes in solid brain structures; the simplest $n\Gamma$ examples here do not extend without more complications to the higher dimensionality networks. Previously (Gregson, 1995), we have explored a lot of properties of the 2D arrays in $(n \times n)\Gamma$ simulations, (but nothing on $(n \times n \times n)\Gamma$).

Qualitative characteristics of the dynamics identified there are often paralleled in CNN theory and simulations. Many of these have been demonstrated in recent years in, for example, the journals *IEEE Transactions on Circuits and Systems* and *International Journal of Bifurcation and Chaos*, though it must be reiterated that usually there is no convincing argument presented that the particular CNNs used are a close analogue of what is needed to explore psychophysics, perception and cognition. To the extent that CNNs are, at the metatheoretical level, a family of processes some generalisations might, however, be useful. It is known that cellular 2D and 3D networks can exhibit pattern formation because they have a spatial frequency instability in the neighbourhood of their equilibrium configuration. Modes of organisation exist in connected sets and, during evolution, a transient state can appear so that unstable modes expand at the expense of stable modes in the dynamics. Oddly, the patterns created in 2D can resemble stripes or spots, checkerboards, or even the surface of the human cortex, but the transition from one pattern to another may be abrupt (Thiran, Crounse, Chua and Hasler, 1995, Crounse, Chua, Thi-

Table 4.6: Cross correlations 6×6 of lags 0 to 4 for Case 1
to compare with Table 4.4; based on a run of 250 points
No cross-coupling, all $e = .25$

above diagonal, n to k, below diagonal reversed k to n

	n to 2	n to 3	n to 4	n to 5	n to 6
lag 0					
-.—	0.156	0.854	0.120	0.794	0.562
0.156	-.—	0.278	0.990	0.302	0.408
0.854	0.120	-.—	0.229	0.994	0.889
0.794	0.562	0.278	-.—	0.253	0.362
0.990	0.302	0.408	0.229	-.—	0.932
0.994	0.889	0.253	0.362	0.932	-.—
lag 1					
-.—	0.152	0.645	0.135	0.586	0.379
0.133	-.—	0.197	-0.803	0.195	0.109
0.589	0.126	-.—	0.195	0.548	0.455
0.531	0.344	0.202	-.—	0.190	0.107
-0.802	0.199	0.114	0.191	-.—	0.453
0.540	0.434	0.193	0.112	0.437	-.—
lag 2					
-.—	0.021	0.467	-0.008	0.404	0.194
-0.023	-.—	-0.044	0.848	-0.038	0.049
0.449	-0.021	-.—	-0.043	0.274	0.135
0.392	0.192	-0.011	-.—	-0.040	0.045
0.852	-0.009	0.053	-0.046	-.—	0.127
0.280	0.145	-0.040	0.029	0.133	-.—
lag 3					
-.—	0.039	0.328	0.028	0.266	0.074
0.065	-.—	0.059	-0.859	0.047	-0.061
0.342	0.067	-.—	0.024	0.121	-0.004
0.294	0.150	0.022	-.—	0.052	-0.051
-0.856	0.013	-0.066	0.063	-.—	-0.017
0.133	0.049	0.017	-0.058	0.024	-.—

Table 4.6 (continued): Cross correlations 6×6 of lags 0 to 4 for Case 1
to compare with Table 4.4; based on a run of 250 points
No cross-coupling, all $e = .25$

above diagonal, n to k, below diagonal reversed k to n

	n to 2	n to 3	n to 4	n to 5	n to 6
lag 4					
-.—	-0.030	0.278	-0.058	0.221	0.043
-0.026	-.—	-0.047	0.841	-0.041	0.052
0.341	-0.015	-.—	-0.068	0.172	0.056
0.304	0.173	-0.035	-.—	-0.038	0.048
0.844	-0.028	0.046	-0.043	-.—	0.063
0.193	0.123	-0.060	0.021	0.111	-.—

Table 4.7: Determinants and roots for 6Γ Case 1

The example in Table 4.6; no λ cross-coupling, $e = .25$

$\{\lambda\}$: $\{1.00, 1.25, 1.29\}$, initial $\{a\}$:$\{3.5, 4.7, 4.1, 4.8, 4.2, 4.5\}$

Y(Re), lags: 0	1	2	3	4
determinants: 4.701	-.0067	2.858	.00258	-1.672

factored to give:

Y(Re) roots: real: 2.2817   .6692   -.2978   conjugate: $-.327 \pm .989i$

Table 4.8: Periodicities of Point Processes

In both cases the $\{a\}$ are $\{3.1, 4.7, 4.1, 4.2, 4.5\}$, run length = 350.
For 6Γ **Case 1** with $e = .25$,

Γ_1	Γ_2	Γ_3	Γ_4	Γ_5	Γ_6
13	8	10	9	12	3
34	90	12	31		31
10	49	28	80		
40		41			

For 6Γ **Case 2** with $\{\lambda\}$ as in Table 4.5,

Γ_1	Γ_2	Γ_3	Γ_4	Γ_5	Γ_6
2	5	2	9	9	9
18	9	14	10	17	18
19	15	18		20	20
28	20	26		28	

ran and Setti, 1996). If the coupled elements in the net are bistable (they have two attractors, like Γ trajectories in Y(Re,Im)) we may get a sort of clustering, and a chaotic distribution of oscillation amplitudes exists in the lattice (Nekorkin, Makarov, Kazantsev and Velarde, 1997). Again, stripes and spots and checkerboards are possible, and phase-locked clusters appear. This phase-locking is a sort of binding.

The analogies with brain action have been noted, in that spatio-temporal structures emerge from cooperation of a large number of apparently disorganised smaller units often themselves operating in a chaotic mode. Not only do we have the emergence of chaos at one level and stability at another, but this coexistence is related to the creation of memory. Ogorzałek, Galias, Dąbrowski and Dąbrowski (1996, p. 1870) call this *spa-*

Figure 4.5: an $(n \times n)$ Cascaded Lattice Process

From the Left Obliquely and Above

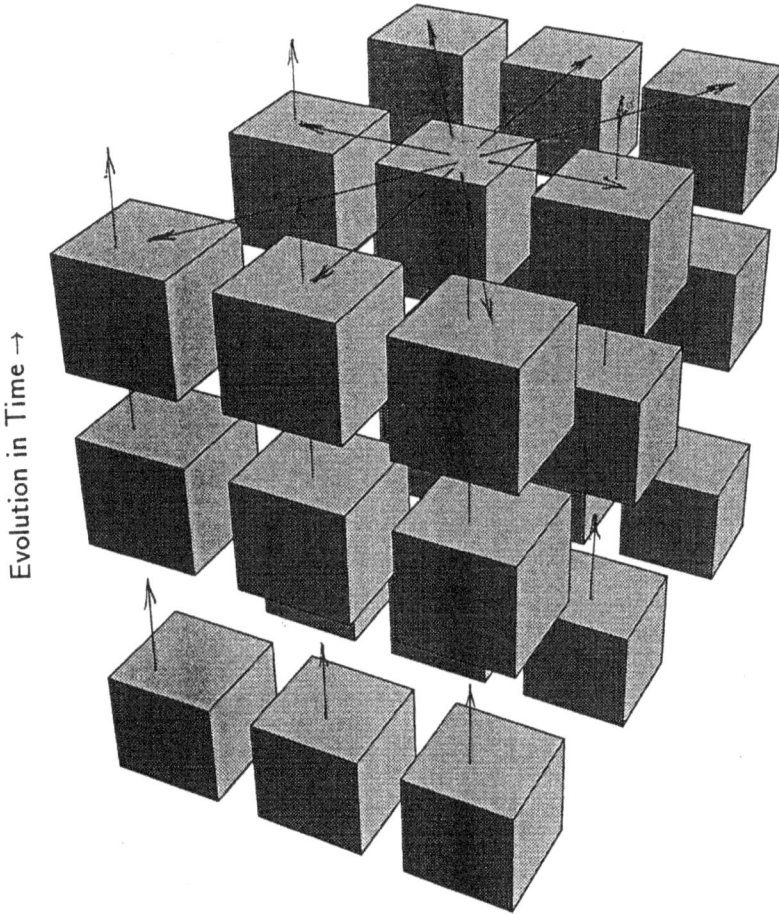

Evolution in Time $\rightarrow$

Diffusion over x, y in space within lattice

tial memory effect, and observe that "Strongly coupled arrays display an interesting phenomena of maintaining information about the position of initialization". Such nets exists in time, not as a frozen representation, and can maintain standing wave patterns which are a symbolic encoding of particular inputs.

Conclusions

The situation with six parallel inputs described is one which is completely deterministic but nonlinear. There are no stochastic components and no residual Gaussian error distributions. But if the outputs are treated as time series, particularly if the series built on the determinants of the 6×6 cross-correlation matrices of lags 0 through 4 is examined, its factoring is only partially informative and can be seriously misleading; it does not give a correct picture of the system's intrinsic dynamics and cannot validly extrapolate in time to predict the future evolution and therefore the stability of the system.

While this point may seem elementary, given the algebraic construction of the system, and the necessary loss in information contingent upon using time series correlational methods, it is reminiscent of attempts to analyse brain mass action by decomposing the EEG in the time or frequency domains. In the case of the brain potential records, we only have the outputs and not the process specification expressed in cascaded nonlinear dynamics, even though the evidence from hippocampal studies suggests that it would be helpful to have that as well, as has been known for some time (Nadel and O'Keefe, 1974, Nadel, 1989).

The mathematics here were developed for modelling psychophysical channels, and not single neurons; the distinction is important, and yet there are some parallels in the dynamics. Koch (1997, p. 209) observes that synapses continually adapt to their inputs in a way that is different from computers that enforce a separation between memory and computation. However, the modern view that neurons have plasticity, feedback, nonlinearities in transmission, and are critically dependent upon the timing of local sequences, is reminiscent of the properties of the cascades reviewed here. Koch (p. 210) says "we sorely require theoretical tools that deal with signal and information processing in cascades of such hybrid analogue-digital computational elements. (there exist). Loops within loops across many temporal and spatial scales. And one has the distinct feeling that we have not yet revealed every layer of the onion."

It transpires that the $n\Gamma$ systems we have studied in depth, for their particular relevance to psychophysics, become a special case of the CNNs with chaotic and bistable elements when they are extended into higher dimensionalities. This inevitably points up the complexities of their potential dynamics, but also helps by suggesting why they can simulate human sensory functioning, and how such functioning is inextricably linked to the higher order functions of memory and cognition.

Appendix 1

Places in Singer's (1993) review on the neurophysiology of oscillating neuronal nets where there are close parallels between the temporal dynamics of the model [4.1, 4.2, 4.4] and what is observed in brain activity are summarised here. Page references are to Singer.

A shift in interest to temporal relations in neuronal processes has begun; because neuronal processes are highly dynamic, temporal codes matter (p. 350).

Oscillatory components in the gamma frequency range (30-60 Hz) are also contained in field potential responses evoked by sensory stimuli (p. 351) which relate to the P 300 and to activity in the thalamic nuclei. The amplitude of the high frequency oscillations in usually small. There have been proposals for a functional significance of the low frequency oscillations in sleep and memory.

There are two sorts of oscillations, pacemakers and the emergent properties of networks. The perceptual functions of the neocortex are based on distributed processes, in parallel and different sites. Because a sensory input, usually visual, elicits a large number of spatially distributed neurons, the binding problem arises; eventually we get one sensory image. The assembly code is relational; "the significance of an individual response depends entirely on the context set by the other members of the assembly" (p. 354). "Temporal coding offers new solutions to segmentation problems that have been notoriously difficult to resolve with amplitude and position codes alone" (p. 355). Cross-correlation studies have actually been based on the analysis of spontaneous activity, which is appropriate for disclosing anatomical connectivity but not for stimulus-induced dynamic coupling.

"Episodes with constant frequency last only 100-300 msecs and can recur several times throughout the response to a continuously moving stimulus." (p. 256). Neither time nor phase are related to stimulus input coordinates (Engel et al, 1991); it is not possible to tell from the responses of individual groups whether they were activated by one coherent stimulus of by two different stimuli. The only clue is provided by the evaluation of synchronicity of the responses of the activated groups. Synchronization by a common subcortical input will not explain binding because it is insufficiently flexible and insufficiently stimulus constellation dependent.

There is a counter argument (p. 360), that "the possibility needs to be considered that synchrony and oscillations are epiphenomena of a system's properties that have evolved for a completely different purpose". Oscillation and synchrony are not necessarily coupled dynamically. Zero-phase lag synchronization can be achieved despite considerable conduction delays and variation of conduction times if the synchronizing connections of the coupled cell groups have a tendency to oscillate (p. 361). The number of assemblies that can coexist in the same cortical region is increased if the oscillation frequencies are variable. because spurious correlations due to aliasing effects will be rare and only of short duration (p. 362).

If Hebbian synaptic modification rules held unqualified, in the long run, the gain of most synaptic connections would be increased leading to unselective and meaningless association of neurons (p. 365). This is the superimposition problem; it can be avoided if the connected neurons have an oscillatory time structure, which is why the n-methyl-d-aspartate (NMDA) mechanisms are of interest [3](Jensen, Idiart and Lisman, 1996)

Appendix 2

The program used with modifications here was originally more complicated, to allow exploration in a wider parameter space. Only one set of a values has been used at a time in the six channel examples here.

Using Fortran 77 programs d6.f for the single mapping (and dl64.f for the bounded cascades), the appended output is generated; there are up to 40 settings of the a_5 values in each block for fixed $\{\lambda\}$, the settings 1 through 40 are increments of a_5 in arbitrary steps, the line reading "lambda 1 through 3" shows the λ settings for the block following.

On each line there are four numbers such as

"1 2.150 2.700 2.500",

these are the setting reference number (which is purely a check reference and has no computational significance) and the values of a_1, a_2 and a_5. Note that throughout this simulation a_1, a_2 are fixed. The variable a_5 implies a_6 moves with it (see table above). Using the relation $e_h = \lambda/(max(a_j)), j \neq h, h, j = 1, ..., 6$ which is defined by **Case 2**, the equivalent e_h are listed only in the cases where

$$\mathbf{C} = |Y_2 - Y_5|(\text{Re}) < .01$$

and as $0 \leq Y(Re) \leq 1$ this is a 1% tolerance bound. The point of listing $e_1, ...e_6$ is to reiterate the pairing of channels 1,2:3,4:5,6; but the independence of e from a. The outputs $\{Y\}(\text{Re})$ are only listed where the constraint **C** is met, to save space in this print-out.

The important effect of the λ settings is to alter where the local matching condition **C** arises. There is one setting for $a_5 \simeq 2.65\ to\ 2.70$ for a matching, but with high λ a secondary output (Y) matching arises with $a_2 = 2.7, a_5 = 3.55$. The precise locality of these solutions to the matching

---

[3] Globus (1995, p. 65) also notes that, for biological realism, the intrinsic modifiability of the transmission properties of NMDA connections has to be considered, in contrast to many parallel distributed processing models in artificial intelligence.

condition will move around as a function of the set of 6Γ parameters. It may be noted that the inequality $|a_1 - a_2|$ is not matched across the $\{\lambda\}$ range by the inequality $|Y_1 - Y_2|$(Re).

Chapter 5

A Bivariate Entropic Analogue of the Schwarzian Derivative

An analogue expression for the Schwarzian derivative of a series had been previously constructed and used to examine the dynamics of a collection of time series from psychological and psychophysiological data. The derivatives of a function in the Schwarzian expression are replaced by information theory expressions based on absolute successive differences of a time series sample. This work is briefly recapitulated, and then the exploratory analysis of the numerical properties of the index is extended to compare the coupling of two time series in parallel, for series that are variously known to be periodic, random, or nearly chaotic. This requires that instead of using the entropy $I^m(x)$ of the mth differences as the building block of the expression, the transmitted information $T^m(xy)$ is used. The manifold of the bivariate form is introduced. Examples of both theoretical and real data from various sources, including psychophysiology, are examined. The possibility of detecting emergent dynamics associated with coupling between series is noted with real examples.

In the sorts of data which are encountered in psychophysiology and psychophysics, there is little room for postulating simple well-behaved dynamics, mathematical tractability has to take second place to empirical plausibility. It is coming to be realised that there are phenomena which cannot be disentangled from noise by employing the statistical methods applicable to long and stationary series, such as Lyapunov spectra, fractal dimensionality, or Volterra filters (Århem, Blomberg & Liljenström, 2000). There are other problems of bias in estimating dimensionality from finite samples (Kitoh, *et al*, 2000), but, in the present context, where comparison of samples from different but perhaps related experimental conditions is of prime interest, second-order bias in any one estimate may not be a serious problem.

In the physical sciences, very long computational series, of the order of 100,000 iterations, may be used to establish with precision the trajectories of, for example, the Henon attractor, throwing away the records of the first 1,000 iterations. But here the opposite problem is faced, the interest lies in the very start of trajectories, before any confidence in the simpler stability of the process may be held.

When interest shifts from single time series to multiple series evolving in parallel then there are additional subtleties to be faced (Cao, Mees & Judd, 1998). The problem of looking for dynamic coupling between short and nonlinear nonstationary series is a popular topic (Schiff et al, 1996; Tass et al, 1998). There are a number of tests for nonlinearity in time series of sufficient length, such as Hinich bicovariance, Tsay's quadratic test, Engle's LM test, which are available on bootstrapped software (Patterson & Ashley, 2000). There is, however, a paucity of procedures for exploratory data screening in the sense of Tukey (1977), which might be employed on very short time series realisations as an augmentation of graphical methods such as recurrence plots, or time scale stretching and variable transformations. That is one reason for the creation of the approach used here; another reason is computational simplicity.

Some investigators hasten to postulate random stochastic processes where they are improbable (Thompson, 1999) and others see only deterministic trajectories onto attractors. Others make some sense of complicated brain dynamics by describing systems that dance between local attractors and random noise at different levels of dynamic organisation (Freeman, 1999, Tirsch *et al*, 2000).

This treatment of the general problem of process identification in time series data was originally motivated by a need to explore, in an initially heuristic fashion, samples of trajectories that might be generated by the Γ dynamics employed in nonlinear psychophysics (NPD), based on maps of a complex cubic polynomial (Gregson, 1988,1992,1995),

$$\Gamma: \quad Y_{j+1} = -a(Y_j - 1)(Y_j - ie)(Y_j + ie), \qquad i = \sqrt{-1}$$

where Y is complex, $0 < Y(Re) < 1$, and $j = 1, ..., \eta$,

Γ can also be written in a succinct matrix form, which elegantly shows that with $e > 0$ (as is usual in substitutions which model real psychophysical phenomena) the process generates negative feedback (Heath, 2000, p. 100). When $e < Y(Im)$ peculiar dynamics evolve (Gregson, 2001), and as $e \to 0$ the linear terms eventually vanish.

The Schwarzian derivative (Schwarz, 1868) was mentioned (Gregson, 1988, p. 27) in the first detailed examination of the Γ dynamics, but its actual computation for realizations of NPD finite-length sample trajectories had not then been pursued. The Schwarzian is defined over continuous functions with derivatives that are stationary in their parameters and may be said to measure a property of expansiveness in which the dynamics fill a local region of their phase space. It is wise not to attempt intuitively to give this idea a common-sense meaning; its importance lies in finding indices which characterize the sort of dynamics underlying the generation of the data, interpreted as time series or as trajectories of an attractor. This problem is fundamental in neurobiology (Arhem, Blomberg & Liljenstrom, 1999) and in nonlinear psychophysics (Gregson, 1995), but in the latter case data are almost always at a coarser resolution of measurement and more intractable.

There is, of course, a diversity of algorithms for identifying the internal dynamics of nonlinear processes when only trajectory samples are observable, earlier methods were collectively reprinted by Hao Bai-Lin (1984), and later methods relying on surrogate generation were collected by Ott,

Sauer and Yorke (1994). If we approach the problems from the perspective of stochastic dynamics (Honerkamp, 1994), which is not really being done here, then there is no novelty in considering information measures. It should be emphasised that the method used here is not in its application restricted to data series which arise in psychophysics. It is neutral with respect to subject matter, but can be employed on series where identification of strong periodicity is meaningless and, at the same time, local autocorrelated bursts of activity may arise but the inter-onset intervals of such bursts may themselves be quasi-random and hence constitute an aperiodic series in time. Series with such characteristics may arise, for example, in the symptomatology of schizophrenia (Ambühl, Dünki and Ciompi, 1962).

It is necessary, wherever possible, to avoid arbitrary and strictly unnecessary constraints on the length of data sets generated from psychological or social processes, because they are usually short, the consequence of transient destabilisation of a process which is already nonlinear, and are far removed from the ideal cases found in theoretical physics. The use of 2^n length samples in frequency domain analyses (Warner, 1998) can have pernicious consequences of process misidentification when nonmetric and/or ubiquitous non-stationarity properties are in fact present.

The Schwarzian derivative

The Schwarzian derivative of a C^3 local diffeomorphism is given by

$$Sf := \frac{f'''}{f'} - \frac{3}{2}\left(\frac{f''}{f'}\right)^2 \qquad [5.1]$$

The Schwarzian of Γ is negative, but only under restrictions on the parameter e (Gregson, 1988, p.28) and the same remains true, by the composition formula, for any monotone branch of a saturated map induced by f_Γ(Graczyk & Świątek, 1998) The form [5.1] is obviously for a single series or trajectory and here it is to be extended to consideration of the linkage between two series evolving in parallel in time, each of which would still have its own form of [5.1]. Van Strein (1988) shows that the Schwarzian is related to cross-ratios and to Moebius transformations, but no such properties are claimed here for the entropic analogue about to be defined.

Coarse Entropy Filtering

For exploratory data analysis of trajectories that might be chaotic or might be almost strictly periodic, Gregson (1999,2000) introduced as a screening procedure the following steps, where the values of the real scalar data y trajectory are bounded, or have been rescaled, into the interval $0 < y < 1$. This initial normalisation removes information about the actual numerical range of the values of the time series, whether or not those numbers strictly satisfy metric properties (as some psychological category scales may not) and makes all values positive. The basis is therefore set for a non-parametric approach. It thus removes information about the first and second moments of the process; that information can be stored separately if it is considered meaningful[1]. This is deliberate, and can have the advantage that series may be compared from various sources when they are originally expressed in units which reflect their different physical origins but are not informative about their relative dynamics.

Here the variable y is real and can be constructed from the complex variable Y in a Γ recursion, as in the example in Table 5.1 used here, by writing $y = polarY(Re, Im)$. There are, of course, two choices if a single variable is to be used from the complex evolution: either only take the real part of Y, or construct the polar modulus $r(Y(Re, Im))$. In the particular context of psychophysics, where it is known that the $Y(Re)$ part resembles most closely the properties of observable response data, this procedure has some justification on those grounds.

The range of y values is partitioned into k exhaustive and non-overlapping ranges, for convenience equal in width, and $k = 10$ is initially sufficient. As a condition of the normalisation, the end categories $(h = 1, k)$

---

[1] It is not necessarily the case that the first and second moments of a non-stationary time series exist; estimates may not converge in the limit (Gregson, 1983).

will be initially not empty[2] Some or all of the remaining $k - 2$ ranges may be empty. Call the width of one such segment $\delta^{(0)}$ (of y); $\delta^{(0)} = .1$ for y but will almost always be much less for $\Delta^m y$, the successive absolute differences of the series. This is the *partitioning constant* of the system. It can be critical for the bivariate form $BESf$ as will be seen later. Call the partitioning constant of the range of the mth differences $\delta^{(m)}$. In the examples used here, $\delta^{(1)}$ is referred to simply as δ in the tables of results, and all $\delta^{(m)}$ are set constant equal to δ. The original partitioning before differencing is given by

$$\forall h \quad \delta_h^{(0)} = k^{-1}$$

if all segments are set equal in width[3] The frequency of observations lying in one segment δ_h, $h = 1, ..., k$ is then n_h. The frequencies are converted into probabilities of segment occupancy p_h, and used to compute the information I_h in that segment. This is metric entropy in the sense of Katok and Hasselblatt (1995, chap. 15).

Summing over all segments gives the total information in the spectrum as

$$I_{\{h\}} = - \sum_{h=1}^{k} p_h log_2(p_h) \qquad [5.2]$$

Then the absolute differences of the series are taken, putting

$$\Delta^1(y_j) = |y_j - y_{j-1}| \qquad [3]$$

which is a discontinuous analogue of differentiation. Further differencing is repeating the operation as in [5.3], so

$$\Delta^2(y_j) = |\Delta^1(y_j) - \Delta^1(y_{j-1})| \qquad [5.4]$$

---

[2] This treatment opens the door to symbolic dynamics, which leads in turn to complexity theory (Adami & Cerf, 2000).

[3] If the segments are unequal but monotonically ordered, this is equivalent to a monotonic transformation of the original y values. It is known that such transformations can affect the autocorrelation of the data series and it is not known a priori for psychological data if the original y necessarily satisfies metric properties; temperatures and arbitrary economic indices, for example, do not.

and this operation can be continued but only until all absolute differences
are zero. Going only as far as Δ^4 is sufficient for empirical purposes. For
each distribution over k segments of Δ^m values a new frequency distribution n_h^m is created, with $\delta^{(1)}$ set by inspection so that the two end partitions $h = 1, k$ are not initially empty for Δ^1 only, and $\forall m, h\ I_h^m$ follows. It
would be possible to introduce two refinements here, which might be crit-
ical. These are, to re-estimate a best $\delta^{(m)}$ at each successive differencing,
instead of leaving all δ the same after estimating $\delta^{(1)}$, and finding by exploration at each differencing stage the value of $\delta^{(m)}$ which maximises I^m.
As shown by an example in Gregson (2000, Table 2), the condition that the
two end partitions $h = 1, k|m$ are non-empty does not necessarily follow
after $m = 1$, when $\delta^{(m)}$ is fixed[4]. If the process has constant absolute first
differences, which is the situation with a random walk with equal increments, as in the familiar drunkard's walk, then I_h^s, $s \geq 1, = 0$. But also,
if the process is strictly periodic, again with constant absolute differences
(which can be obtained if the sampling frequency is simply related to the
process frequency) then I_h^m vanishes for $m = 2$.r Some series which are encoded in arbitrary units y may be monotonically transformed to $y^* = f(y)$
so that their absolute first differences become constant[5].

A Schwarzian derivative analogue

Suppose that we substitute within [5.1] I^m values for f, so that for example
I^3 corresponds to f''', employing I^m summed over the whole range of k
segments. From an example (Gregson, 1999, 2000) using three series of
length 280, one of them generated by the convolution of two Γ series (each
of which is in a complex variable, Y) with the output variable expressed
as $y = polarY$, compared with a strictly periodic series, and a random

---

[4] This, when it happens, is the consequence of ESf not being the analogue of the C^3
diffeomorphism condition which defines the Schwarzian.

[5] Such transformations to or from this simplest structure can affect the autocorrelation
spectrum and produce or delete apparent evidence of long-term memory in the process
(Heyde, 2000). Mandelbrot (reviewed by Feder, 1988) had shown that long memory process can have a fractal structure.

series, we may construct the required variable. The values of I^m are given in Table 5.1, ignoring the minus sign in [5.2]. This is to create an entropy analogue of a Schwarzian derivative. To give this a name, call it ESf. It is expected that ESf will be negative if a type of dynamic stability is present in the process.

Table 5.1
Values of I^m for $k = 10$ on three series

Type of Series	I^1	I^2	I^3	I^4
convoluted Γ	2.3604	1.6055	2.1345	1.4164
Periodic	1.000	0.0133	0.0434	0.0434
Random	2.9747	2.499	2.3062	2.1183

Then by definition ESf takes the values

$$ESf := \frac{I^3}{I^1} - \frac{3}{2}\left(\frac{I^2}{I^1}\right)^2 \qquad [5.5]$$

Substituting in [5.5] gives:
$\quad ES\Gamma = .1571, ESPeriodic = .0431, ESRandom = -.4254$

Table 5.2 shows some results from applying [5.5] to various Real and Imaginary trajectories of $Gamma$ to investigate the behaviour of the dynamics as a function of the parameter product ae in the near-chaos region, where after about 50 or more iterations the process explodes. Estimates of the Hurst exponent $0 < H < 1$ and the fractal dimensionality $1 < D < 2$ have been given, derived from an algorithm used by West et al (2000) on what they call random fractal time series. By a standard result, $H = 2 - D$ in the 2d plane. For $0 < H < .5$ the process is anti-persistent, for $H = .5$ it is effectively random, and for $.5 < H < 1$ it exhibits persistence. The algorithm sometimes fails on short series; the values where given make sense but are numerically not necessarily accurate, nor very meaningful on short unstable realisations such as trajectories **onto** an attractor, as opposed to being **on** the attractor.

Table 5.2: D,H,ESf, and Gamma trajectories
$\eta = 50$ in all cases examined

Re/Im	a	e	D	H	ESf	ae	period
Re	2.70	.4212	1.7506	.2494	**-.2353**	1.137	2
Im	2.70	.4212	1.6979	.3021	.3076		4
Re	3.00	.3830	1.3464	.6536	**.0637**	1.149	2,1;3
Im	3.00	.3830	1.1971	.8083	.1397		3
Re	3.25	.3100	1.4403	.5597	**.0637**	1.007	6,1;7
Im	3.25	.3100	1.3213	.6787	-.8016		7
Re	3.55	.2840	[< 1]*	[> 1]*	**.0637**	1.008	3,3;6
Im	3.55	.2840	1.1391	.8609	.1648		6
Re	3.85	.24177	1.7375	.2625	**.2677**	.931	3,1,1,1;6
Im	3.85	.24177	1.7325	.2675	-.8016		6
Re	4.00	.2300	1.8908	.1092	**.1648**	.920	2
Im	4.00	.2300	1.8526	.1474	.1672		2
Re	4.25	.2160	1.9270	.0730	**.0637**	.918	2,1;3
Im	4.25	.2160	1.9718	.0282	.1397		9
Re	4.45	.1500	1.6982	.3018	**.0206**	.668	1
Im	4.45	.1500	1.7383	.2617	.0456		4
Re	4.73	.1300	1.9719	.0281	**.1648**	.615	2
Im	4.73	.1300	1.9704	.0296	.3703		2

* these estimations are erroneous and almost certainly meaningless; the algorithm has failed with this time series.
The notation for periodicity indicates repetitions of identical numerical values within a cycle, where they occur. The final number is the periodicity.

The ESf values must be understood as each having a confidence interval associated with them, they are not constants. Finding what that confidence interval is might be a suitable problem for MCMC analyses. The positive value of ESf for convoluted Γ is not necessarily the same as a negative value for unconvoluted Γ. Thus, convolution can change the stability indexed by the Schwarzian analogue. In comparing different data sets, both the between and within-sample variabilities of ESf should be considered.

Extending to the bivariate case

The formula of [5.5] is now extended to consideration of two series in parallel, where one may be the same as or different from the other, and one may be lagged on the other. Without loss of generality, this also covers the case where one leads on the other. This problem is fundamental in psychophysiology, where nonlinear coupling between the dynamics of two or more attractors is found in biological systems (Schiff et al, 1996; Breakspear, 2000).

In information theory, by definition, for two series $x : g = 1, .., k$ and $y : h = 1, .., k$, the transmitted information $T(xy)$ involves $I(x,y)$ on the $n \times n$ contingency matrix derived from [5.2] by writing

$$I(xy) = I\{g, h\} = - \sum_{g=1}^{k} \sum_{h=1}^{k} p_{g,h} log_2(p_{g,h}) \qquad [5.6]$$

which gives

$$T(xy) = I(x) + I(y) - I(x, y) \qquad [5.7]$$

so [5.5] becomes

$$BESf := \frac{T^3}{T^1} - \frac{3}{2}\left(\frac{T^2}{T^1}\right)^2 \qquad [5.8]$$

There are two computational steps which must be decided before calculation of ESf and $BESf$; the magnitude of δ, the partitioning constant has to be set to maximise the I^1 for the series with the greatest variability of first differences, and the choice of lags. The original normalisation has already operated on I^0. In these simple examples, both have been done by inspection and trial and error. If the series examined are strongly periodic, as in sinusoidal forms, then the lags may be set to detect the dominant frequency of the cross-coupled series. The problem of phase synchrony between neural sites has also previously been attacked by using entropy measures (Tass et al, 1998).

As any series which is not on a point attractor varies as it is lagged on time in terms of ESf, lagging one of a pair of series against the other yields a range of ESf values for one against a fixed ESf for the other, as is illustrated for two realizations of a random series lagged on itself, of length 200, in Table 5.3. The series was created using a Fortran 77 program from the Numerical Recipes library (Press, Flannery, Teukolsky & Vetterling, 1986). It is seen that, for random series, $BESf$ can have very unstable outlying values, which are associated with high $T(xy)$, and can fluctuate in its sign, as a function of the partitioning constant δ (of x and y), which is an arbitrary parameter of the derivative. Tables 5.3 and 5.4 are constructed to illustrate this point. It is, therefore, suggested that a minimum variance solution of $BESf$ over a set of lags should be sought. It is seen in Table 5.4 that all the signs of $BESf$ are consistently negative and thus also compatible with the signs of component ESf terms.

The range of values under $ESf(1)$ in Tables 5.3 through 5.8 arise from taking successive rectangular moving window samples of the process and may be averaged, if the process is taken as stationary, to get a mean estimate.

The next case, in Tables 5.5 and 5.6, uses some EEG data from a study by Dr. Kerry Leahan at The Australian National University. These are the first examples of real data in this chapter. They show signs of near-chaos and some dominant frequencies. They had been separately analysed using frequency spectra and by estimating the largest Lyapunov exponent. The two series are drawn from different sessions and are not, therefore, synchronous or phase-locked. In a situation with a cranial montage of EEG recordings synchrony would be available. Table 5.6 is the preferred solu-

Table 5.3

Values of ESf and $BESf$ on a lagged random series
partitioning constant = .04

lag	ESf(1)	ESf(2)	BESf(1,2)
0	-.60339	-.60339	.1371
1	-.61201	-.60339	-.1067
2	-.60982	-.60339	.0021
3	-.61193	-.60339	-.8043
4	-.60849	-.60339	.7013
5	-.61339	-.60339	-.3997
6	-.61565	-.60339	.2106
7	-.62309	-.60339	.0531
8	-.63860	-.60339	-.6176
9	-.63388	-.60339	-.0730
10	-.62859	-.60339	-30.0559

Table 5.4

Values of ESf and $BESf$ on a lagged random series
partitioning constant = .038

lag	ESf(1)	ESf(2)	BESf(1,2)
0	-.63087	-.63087	-1.6144
1	-.63989	-.63087	-1.2194
2	-.63779	-.63087	-.7811
3	-.64043	-.63087	-5.5540
4	-.63441	-.63087	-3.9493
5	-.63874	-.63087	-1.9970
6	-.64264	-.63087	-1.4932
7	-.65229	-.63087	-.4745
8	-.67449	-.63087	-.0096
9	-.67168	-.63087	-.1700
10	-.66082	-.63087	-1.0143

tion.

Another pair of EEG series is presented in Table 5.12, that has been drawn from a real library *www.physionet.org* example, which is a synchronous pair but about which no physiological details concerning location and experimental conditions are known. As those details are missing, it is relegated to the Appendix merely for illustration of the instability that $BESf$ can have.

The ESf values are of comparable magnitude to those given in Table 5.5, the D and H values were computed as in Table 5.2 for Γ series.

Table 5.5
Values of ESf and $BESf$ on two EEG
series under relaxation conditions
partitioning constant = .07

lag	ESf(1)	ESf(2)	BESf(1,2)
0	-.66070	-.54500	.0388
1	-.65467	-.54500	.0512
2	-.65213	-.54500	-.7596
3	-.64850	-.54500	-.1207
4	-.65046	-.54500	-.7527
5	-.64447	-.54500	.2655
6	-.64480	-.54500	-.0816
7	-.66602	-.54500	-.6390
8	-.68870	-.54500	-1.3849
9	-.70055	-.54500	-.6579
10	-.73027	-.54500	.4720

The third example, in Tables 5.7 and 5.8, is created by using two strictly periodic series

$$x_{1,j} = a_1 \cdot sin(b_1 \cdot j) + a_2 \cdot cos(b_2 \cdot j) \qquad [5.9]$$

$$x_{2,j} = a_2 \cdot cos(b_2 \cdot j) \qquad [5.10]$$

Table 5.6

Values of ESf and $BESf$ on two EEG
series under relaxation conditions
partitioning constant = .10

lag	ESf(1)	ESf(2)	BESf(1,2)
0	-.73528	-.48770	-.1226
1	-.72137	-.48770	-.3159
2	-.70651	-.48770	-.3454
3	-.70323	-.48770	-.1376
4	-.70651	-.48770	-.4096
5	-.69632	-.48770	-.0565
6	-.69594	-.48770	-.0500
7	-.71949	-.48770	-.0725
8	-.74613	-.48770	-.3844
9	-.76490	-.48770	-.1546
10	-.79342	-.48770	-.4595

where $a_1 = 1.0, a_2 = .5$, and $b_1 = .7855, b_2 = .4$. The periodicity creates marked variation in $BESf$ as a function of the lags used. In effect we are here sampling from a series using a moving rectangular window of fixed length. It will be noted that some of the ESf values are now positive, as was derived in a case from Table 5.1. Table 5.7 is the preferred solution, but this is not the result of an exhaustive search, merely illustrative. It appears that cross-linkage between sinusoidal series, which are the easiest form of strictly periodic series to create (unless we consider Walsh functions), generates wild outliers in $BESf$. With the possible exception of some experiments in psychoacoustics using sinusoidal forcing functions, this example is really not very plausible for real data. The distribution of $BESf$ obtainable by Monte Carlo methods would need lag series many times longer that the illustrative tables presented here.

Table 5.7

Values of ESf and $BESf$ on two
sinusoidal series, [5.9] and [5.10]
partitioning constant = .079

lag	ESf(1)	ESf(2)	BESf(1,2)
0	-.39387	.13742	-5.7812
1	-.36826	.13742	-.4466
2	-.34948	.13742	-4.0372
3	-.28921	.13742	-1.1511
4	-.25154	.13742	-1.1320
5	-.17656	.13742	-6.0086
6	-.14426	.13742	-.5530
7	-.06312	.13742	-1.9803
8	-.00925	.13742	-1.5330
9	.04213	.13742	-.3300
10	.08161	.13742	-4.8755

Table 5.8

Values of ESf and $BESf$ on two
sinusoidal series, [5.9] and [5.10]
partitioning constant = .08

lag	ESf(1)	ESf(2)	BESf(1,2)
0	-.28765	.13260	-.9329
1	-.26615	.13260	-303.5240
2	-.24998	.13260	-146.9375
3	-.19211	.13260	-1.0289
4	-.15645	.13260	-.8354
5	-.08514	.13260	-10.5506
6	-.05563	.13260	-7.4854
7	.02019	.13260	-.1838
8	.06890	.13260	-.0000
9	.11239	.13260	-.3086
10	.14464	.13260	-1.1079

The manifold of the lagged BESf

If there exist two series in temporal synchrony, and each may be lagged upon itself and upon the other, then four $BESf$ can be computed; this set we will call the *manifold* of the $BESf$ up to lag ℓ. It may be seen that this manifold is a non-parametric form of the set made by computing the autocorrelation spectra and cross-correlation spectra of two series each with metric properties.

For illustration, Table 5.9 shown the manifold of the examples in Tables 5.7 and 5.8. One column is therefore repeated from Table 5.7.

Table 5.9
The manifold of $BESf$ for $\ell = 0, .., 10$ on two
sinusoidal series, [5.9] and [5.10]
partitioning constant = .079

lag	BESf(1,1)	BESf(2,2)	BESf(1,2)	BESf(2,1)
0	.1374	-1.8582	-5.7812	-.7466
1	-.0240	-3.7648	-.4466	-53.4660
2	-.1504	-.3957	-4.0372	-43.3347
3	-.1952	-2159.8242	-1.1511	-.9275
4	-.1948	-6540.9619	-1.1320	-.8152
5	-.0291	-.3353	-6.0086	-4.0959
6	-.0468	-3.3587	-.5530	-3.8287
7	-.1021	-.6232	-1.9803	-.6990
8	-.0195	-.4921	-1.5330	-.7392
9	-.3657	-1.4399	-.3300	-16.6598
10	-.0637	-.1693	-4.8755	-85.2626

In these tables, the lag zero $BESf$ values are, in effect, ESf values. It is seen that the manifold incorporates the effects of both leads and lags, from $BESf(x,y)$ and $BESf(y,x)$ compared, and that simpler sinusoids can produce $BESf(x,x)$ which are degenerate. For data exploration it may be preferable to compute $log(BESf)$ values, it is the sequential pattern over ℓ, which is informative.

The next real example is one which produced a surprise; it shows that this method, though crude, can have some heuristic value, in that it leads us to ask different questions about how the data are generated.

Recordings of spike trains per second in the visual cortex of an anaesthetised cat were recorded by two independent tungsten-in-glass microelectrodes with tips about 0.5mm apart. Each stimulus induces a brief time series in the two neurons being recorded; call each such train an epoch. The onset times of the two parallel series in an epoch are synchronised; there are no lag effects shown in Table 5.10. There are other parallel response series under a control condition which are not examined here.

Table 5.10 gives the values of ESf and $BESf$ for the first 33 epochs. Though the original data analysis in terms of correlations (not done by me) suggested that the cells were not coupled, the pattern in Table 5.10 suggests that they are in terms of some weak dynamics. Such weak synchronization during conscious perception has been reported by Srinivasan, Russell, Edelman and Tonini (1999). This can be checked by using surrogate series methods (Theiler et al, 1992), but there are reported statistical difficulties with such methods. In dynamical terms, this suggests an hierarchical nesting of recursive maps. Dynamics in brain activity, arising as a consequence of stimulation as complicated as this have been reported by Freeman (1999). He suggests that input from receptors increases the strength of interactions between neurons, so that a loss of degrees of freedom is entailed.

Discussion

There are many ways in which the parallel dynamics of two evolving processes can be coupled, even in the simplest cases which presuppose stationarity both in the component processes themselves and in the cross-linkage. For example, the resultant series may be a weighted average, a vector sum, a convolution, or, as here, a transmission of information.

Table 5.10: Cat visual cortex V1 epochs

Epoch	N	$ESf(1)$	$ESf(2)$	$BESf(1,2)$
1	60	-.4632	-.5421	-1.2157
2	115	-.7450	-.6601	-2.4115
3	60	-.9041	-1.1539	-3.5910
4	60	-.6069	-.8327	-2.0553
5	40	-1.2018	-.9096	-1.4873
6	40	-.2575	-.2011	-.4459
7	40	-.6027	-.5572	-1.8936
8	30	-.6307	-.7876	-1.0410
9	35	-.3268	-.3320	-2.4721
10	35	-.3232	-.3959	-1.5423
11	35	-.4132	-.5221	-4.6697
12	60	-1.4888	-.9453	-1.3293
13	30	-.2093	-.2617	-.8109
14	50	-.9420	-.8911	-.2552
15	40	-.5771	-.6522	-7.0735
16	30	-.5497	-4820	-1.0638
17	45	-.9888	-.8400	-1.4512
18	40	-.2526	-.2060	+.1104
19	40	-.2816	-.3953	-2.6637
20	60	-.4522	-.5129	-.2042
21	60	-1.0393	-.9636	-1.8879
22	40	-.9117	-.6467	-1.5392
23	35	-.5436	-.5366	-1.3955
24	50	-.2620	-.2982	+.0266
25	30	-.4151	-.3652	-.7299
26	30	-.6285	-.6636	-8.9292
27	55	-.7805	-.6349	-.4545
28	50	-.3448	-.3656	-.6507
29	50	-.6630	-.8119	-1.2003
30	40	-.6088	-.6158	-1.1806
31	20	-.4625	-.5934	-2.8674
32	30	-.6346	-.6642	2.1529
33	40	-.2985	-.3468	+.9147

In this table N is the effective time series length within the epoch as analysed.

The index $BESf$ is, from our tentative exploration of its properties, not particularly easy to interpret. It would seem wise always to preserve the two ESf indices as well, as has been done in the Tables, and to compute the manifold if little is known a priori about the dynamics involved. If large reference samples are created against which any empirical realisation may be compared, then the type of the dynamics involved can potentially be identified. In this sense, ESf resembles (but is not reducible to) the Hurst exponent of a time series (Feder, 1988). However, the original meaning of the Schwarzian as a measure of expansiveness is probably lost, a better intuitive interpretation is one of information impoverishment in a restricted sense.

There are two reasons why computation of ESf and $BESf$ may be useful: to identify the dynamic signature of the process, just as the Lyapunov exponents are used in chaotic dynamics, or to match against a given reference sample in order to test for stationarity in the outputs of a process, even if the underlying dynamics of that process are not fully understood. It is critically important that the phase lock is known, if two series are synchronous, and that the increment (j in [5.9], [5.10] for example) is the same, or a known multiple in one series of the other, in real time units for two or more series compared.

The comments raised here do resemble observations of Tass et al (1998), in particular, the need to try to circumvent the "hardly solvable dilemma 'noise versus chaos' irrespective of the origin of the observed signals" (op cit. p.3293). It is now known that phase synchronisation of chaotic attractors can arise or be induced (Chua, 1999) and a sort of phase locking may be observed. The amplitudes of synchronised systems can remain chaotic and affect the phase dynamics in a way that also arises as a consequence of the presence of external noise. Local phase slips may be observed, which means over a long series the question of synchronicity is irresolvable. Tass et al (1998) make the important point that synchronisation is not the same thing as cross-correlation, so instead of using the algebra of linear time series statistics, they proposed two measures, one of which, like ESf and $BESf$, is based on entropy.

The physiological data we have seen so far and been able to anal-

yse, from the cat study and now from some data provided by the
www.physionet.org site, whose exact cortical source is not given, both
indicate that BESf might show long periodicities. Those long periodicities seem to emerge uncorrelated with the observed structure of the ESf
sequences. There is some evidence that binding between closely adjacent neural locations is mediated by 40Hz activity (Llinas & Ribary, 1993;
Singer & Gray, 1995) in which case, sampling at 300Hz for ESf series would
show 40hz as a slower quasi-periodicity in BESf. This fundamentally important matter needs more data for its resolution.

One advantage we would claim for our approach here is that relatively
short series can be explored. This is important when the non-stationarity
of psychological data is so often apparent and the need to compare short
epochs associated with specific known or suspected stimulus intervention
arises. Schiff et al (1996) note the necessity of identifying the direction of
causality in bivariate linkages and also confirm that systems with nonlinear cross-correlation will show mutual nonlinear prediction when standard analyses with linear cross-correlation fail. This matter has also received comment from Vinje and Gallant (2000). To revert to the nonlinear
psychophysics which first motivated this study, the series we observe are
interpreted as trajectories and the coupling between them is dissipative.
The theory of such coupling involves what is called 'stochastic synchronisation', which is a topological concept assuming a mapping up to a diffeomorphism between nonlinear attractors, but its identification so far necessitates using delay coordinates (a standard method in the identification of
chaotic dynamics), which demands longer series than in the examples we
realistically can obtain in psychophysics.

Note

The programs used here for ESf and $BESf$ were written by the author using Fort77 in a Linux environment. I am indebted to Kerry Leahan for the use of some of her EEG data, to Dr Mike Calford for data
from psychobiological research, to the Senior Vice President of the Templeton Group for providing me with an economic time series and to Michael

Smithson for pointing out the partial resemblance to the Hurst index. This weak resemblance is illustrated in the example from economic data in the appended table.

Appendix: Bernstein economic data and Physionet data

Table 5.11 shows an analysis of a series of monthly values of an economic index which was devised by a New York investment consultancy, Bernstein, in the form of integer values that range from -7 to +10, reflecting the under- or over-valuation of stock. It was used for illustration by the Templeton Global Growth Fund in its 1999 annual report, as a histogram graph.

The ESf, fractal dimension D, and Hurst index H for each subseries of 90 days are at the foot of the table, below the autocorrelation spectra. The autocorrelation coefficients are carried through to 13 lags because the data are monthly and some annual cycling might be in evidence around lag 12. It is not, however. These subseries are too short for ideal computation of the various indices, but the whole series is very non-stationary and non-linear.

The interest lies in the increasing instability of the dynamics as late 1999 is approached. Using ESf it is seen that the increasing unpredictability is continuous from 1970 onwards, whereas using H the anti-persistence increases only in the last epoch. Note that the series becomes relatively unstable in the last column; this is what economists call 'volatility', which appears here to be a mix of high variance and antipersistence. In terms of giving advice to investors, it means that fund managers are, unsurprisingly, less confident in making predictions when volatility is high.

Each of ESf, D and H are filters. Each preserves and destroys information in an almost unique way. To get more insight into the evolving dynamics it can be expedient to use all three, and even more indices. One index suggests that there might be an abrupt change in the dynamics, a jump into volatility, another suggests that the rot had insidiously begun quite some time previously. As used here, all three filters smudge out the dynamics within data blocks, so that to search for a hypothetical sharp locus of a dynamical change at one month other filters would be needed.

Table 5.11 The Bernstein-Templeton Economic Series
The four successive periods are 90, 90, 90 and 89 months long respectively.

Period;	Feb 1970→	Aug 1977→	Feb 1985→	Aug 1992→
mean	-.389	-.044	-.633	.000
s.d.	1.590	1.763	1.792	2.152
min	-4	-7	-7	-5
max	+4	+4	+3	+10
$ar1$	-.0481	.0057	.0763	.5024
$ar2$	.0180	-.1262	-.1570	.1481
$ar3$	-.0223	-.0008	.0596	.0194
$ar4$	-.1248	-.1293	.0013	.0850
$ar5$	.0170	.0424	.0584	.1384
$ar6$	.0627	.0421	.0280	.0388
$ar7$	-.1342	-.0116	-.0326	-.1238
$ar8$	.0357	-.2510	-.0015	-.1893
$ar9$	-.0141	-.1511	-.0800	.0461
$ar10$	-.0675	-.0745	.0697	.0704
$ar11$	-.1175	.1140	.2486	.0874
$ar12$	-.0735	-.0080	.0760	.0558
$ar13$	-.1401	-.0834	.0063	.0049
I_1	2.1474	2.0751	1.7980	2.1197
I_2	1.9214	1.9260	1.8279	1.9474
I_3	1.9421	1.9840	1.7050	1.3642
ESf	-.2965	-.3361	-.6019	-.6225
D	1.3996	1.4354	1.4235	1.1015
H	.6004	.5646	.5765	.8985

Table 5.12: $D, ESf, BESf$ **in a Physionet EEG series**
3 msecs time intervals, recordings in μvolts

segment	D(2)	ESf(1)	ESf(2)	BESf(1,2)	BESf(1,2,1)
1-100	1.9121	-.4607	-.4225	-1.3548	+.4376
101-200	1.7592	-.4577	-.3633	-1.1785	-.5894
210-300	1.7469	-.2760	-.3588	-.2902	+.1218
301-400	-	-.4715	-.5906	-.4803	-2.2851
401-500	-	-.6452	-.5974	-2.9476	-1.5224
501-600	-	-.7287	-.5790	-.3100	-.1625
601-700	-	-.4462	-.5339	-.4956	-.6827
701-800	-	-.4621	-.4997	-2.3512	-.1527
801-900	-	-.6054	-.4259	+.9309	-4.1840
901-1000	-	-.4133	-.3916	-.0529	-.0490
1001-1100	1.7882	-.3611	-.3402	-2.3839	-.3468
1101-1200	1.8602	-.5401	-.5729	-.4712	-1.9269
1201-1300	1.7737	-.5171	-.4922	-1.2113	-.2312
1301-1400	-	-.3791	-.4328	-.1580	-.1962
1401-1500	-	-.2443	-.2619	+.0242	-.1154
1501-1600	-	-.8589	-.4347	-.5036	-1.7437
1601-1700	-	-.4356	-.4266	-.1498	+.0114
1701-1800	-	-.3860	-.5122	+.0733	-.5679

D(2) is the estimated fractal dimensionality of the second series. It has only been computed where it showed possible instability.
BESf(1,2,1) means the cross-Schwarzian with the second series lagged by one step.

Chapter 6

Tribonacci and Long Memory

There are various ways of constructing systems that jump in their dynamics from one configuration to another. Systems that move from edge-of-chaos into a sort of saturated stability as their complexity increases are considered to be a basis for a new sort of thermodynamics where entropy is not always increasing, but the complexity of both a biological system and its connected potential environment are increasing together (Kauffman, 2000). This chapter uses a completely artificial model whose properties can at least heuristically provide illustrations of qualitative jumps in dynamics, some of which are irreversible, unlike transitions through cusp catastrophes, and identify some of the results of relaxing purely deterministic dynamics into partly stochastic constructions. Identifiability is again a central theme of our explorations. As this is a complicated topic a short preamble of the chapter follows.

Time series processes which can be derived from the Tribonacci series by damping its explosive properties are examined. A modification of the Tribonacci series formed by imposing a modular constraint on its expansion, called Tribmodk, is used to illustrate the effects of small changes in its parameters on a diversity of statistical properties variously used to characterise a quasi-periodic trajectory, which may be suspected to be chaotic or at the edge-of-chaos and almost certainly falls into the family of transient nonlinear nonstationary series that arise in psychology. The entropic analogue of the Schwarzian derivative, ESf, is again used, in first-order and higher-order analyses, and the potential use of the Tribmodk series as a source of time series in psychophysical serial extrapolation experiments is noted. These series create conceptual problems in distinguishing in human sequential performance between identification, estimation and prediction. Comparative analyses, falsely assuming that the process is purely stochastic and stationary, are tabled to show that the process can be described but not functionally mapped by excessively complicated classical ARMA modelling. Mixing deterministic and stochastic components has complicated consequences, not necessarily diminishing the stability of the trajectories.

There are two main ways in which one might want to try and explore the relation between an observed data string made up of human responses and the nonlinear process that leads to its observable evolution through some sequences of filtering: one is to create a process that is known fully in its algebra to be nonlinear, use it as a stimulus generator, and see what responses it evokes in sequence over time; the other is to start with properties of the observable response sequence and conjecture what nonlinear dynamics might have produced it. The first is found in the few experimental studies that have used series extrapolation as a dynamic response task, and the second is to start with simple series, such as point processes, and use a known nonlinear attractor to produce the inter-stimulus response interval statistics. Both can be found in the relevant literature; the second (Han, Kim & Kook, 2002) is mathematically better understood but is psychophysically useless because it relies too readily on the generation of long series under stationarity.

In psychophysiological data there are various sorts of sequences: response that vary in magnitude with the same time intervals between them; series of time intervals with the same point processes (of unit magnitude) that are recorded; and series that involve both at the same time variable intervals between events and variable magnitude events. The third sort are, in principle, the most interesting.

A fundamental difficulty lies in determining what is a close copy or an extension of the trajectory of a dynamical process. Given some series, it is not impossible for the human observer, by a sort of dynamical extrapolation, to create a few further terms which match it in the first two moments; but it is not expected that the full autocorrelation spectra of the process could be matched for something like an ARMA process. If the generator of the series is nearly chaotic, then the observable output may have embedded in it local short epochs that are quite predictable, separated by epochs that are indistinguishable from random evolution (Geake & Gregson, 1999). The random parts are predictable in their first two moments and effectively stationary therein; the recurrent short arpeggio-like patterns are identical copies of one another but their onsets are not predictable by simple extrapolation. The use of the Lyapunov spectra to characterise a chaotic sequence rests on the property that prediction is locally possible but falls off exponentially with extrapolation into the future. Predictability of mixed chaotic and stochastic processes may be viable if the resolution of prediction of individual terms is only with a predetermined limit $\pm\epsilon$. What is predictable mathematically in a simulation exercise is not the same as prediction performance by human observers, but the first may be used as a baseline measure of prediction efficiency.

If it is seriously contended that the process by which a human observer creates series of responses without feedback is analogous to the tracing out of an attractor trajectory on the invariant manifold, then the Poincaré sections (plotted as re-entry maps) of the theory and data strings may be generated and compared under stationarity. It does not necessarily follow that the prior creation of a stimulus series can overwrite the internal dynamics of response generation once the stimulus series is discontinued. There may be two different attractors involved.

Psychological models of perception and memory usually make provision for three boundary conditions, a short-term memory of limited capacity feeding into long-term memory of almost unbounded capacity, and an upper bound on the range of admissible sensory inputs. There are other memory systems now experimentally identified (Baddeley & Logie, 1999), but this distinction between transient working memory and the storage of retrievable memories of various sorts is generally accepted. For a human observer to observe U_{n-1}, U_{n-2} and U_{n-3} and then respond with U_n requires holding previous terms in memory. Three terms is probably within the capacity of short-term working memory. Modelling of this may be done in a diversity of mathematical ways, here one of the simplest forms of such a process model is used to illustrate how time series with nonlinear properties result from the imposition of the boundary conditions. The subsequent analysis of the identifiable statistical properties of the series is explored with a diversity of measures.

The Tribonacci Series

A recurrence relation, which is sometimes called the generator of the Tribonacci series, by a humorous analogy with the much better known Fibonacci series, is

$$U_n = U_{n-1} + U_{n-2} + U_{n-3} \qquad [6.1]$$

where U_n usually is taken to have only real positive integer values, and has as its equation for deriving its roots as

$$x^3 - x^2 - x - 1 = 0 \qquad [6.2]$$

Unlike the Fibonacci series [6.2] has only one real root, which is

$$\eta = 1.83928675521416...$$

and is not transcendental.

Cundy (2000) notes that η can be given an explicit value, which is

$$\eta = \frac{1}{3}[1 + (19 + 3\sqrt{33})^{1/3} + (19 - 3\sqrt{33})^{1/3}] \qquad [6.3]$$

and he lists some curious and useful properties of η, such as

$$(\eta - 1)(\eta^2 - 1) = (\eta + 2)(\eta - 1)^2 = 2 \qquad [6.4]$$

or

$$\eta^4 + 1 = 2\eta^3, \quad \text{and thus } \eta + 1/\eta^3 = 2 \qquad [6.5]$$

Here an attempt is made to examine some series derived from [6.1] which do not explode and hence might be of interest as models of stable time series with long internal memories but a limited capacity to expend energy or to store outputs. As Cundy (2000) notes, [6.1] is a truncated case of the general equation

$$x^n = x^{n-1} + x^{n-2} + ... + x + 1 = (x^n - 1)/(x - 1) \qquad [6.6]$$

which has a solution $x = 2 - 1/x^n$, that obviously approaches 2 steadily with increasing n. We may also comment that as a process [6] is not stationary in its order for the first $(n - 1)$ terms.

It is apparent that n in [6.6] is an assumption about the short-term memory of the system, and is taken to be a constant independent of U. Let us call this the order of the memory and denote it by m, to leave n for use as a general suffix. In real experimental memory data, the capacity of the component memories is not observed to be constant, and m could therefore be better treated like a variable. This does not affect the analyses in Tables 6.1 and 6.2.

Let us rewrite [6.1] in two different ways and explore each separately. First, dampen the cumulative nature of [6.1] by rescaling[1] any term $U_n >$ 10 by $mod10$. So, if $U_1 = U_2 = U_3 = 1$, to start the recurrence, then $U_4 = 3$, $U_5 = 5$, $U_6 = 9$, $U_7 = 17.0(mod10) = 1.70$, $U_8 = 15.7(mod10) = 1.57$, $U_9 = 12.27(mod10) = 1.227$, $U_{10} = 4.497$ and so on. Rewriting [6.1] we now have

$$U_n = \begin{cases} U_{n-1} + U_{n-2} + U_{n-3} & \text{if } U_n \le k \\ k^{-1}U_n & \text{if } U_n > k \end{cases} \qquad [6.7]$$

where the role of k is to put an upper bound on responses to inputs, so that the process does not explode. The need for such a bound has been recognised since Herbart in 1812 (cited by Gregson, 1988).

---

[1] This is not $modk$ in the proper sense, if it were, then, for example, $U_n = 17(mod10)$ would become $U_n = 7$. If that definition of mod is used, then the series takes only integer values and could be represented as a Markov chain over k states.

The important point to grasp is that this series is quasi-periodic[2]. Running the first 200 iterations to three decimal places suggests that it is approximately period 19; as 19 is a prime number, this is a curious property, not one associated with the more familiar bifurcation series or period three recurrences in chaotic dynamics. It then raises the question, does it look like a series at the edge of chaos? In general, the series [6.1] can be rewritten $mod(k)$, so to give it a name let us call it **Tribmodk**. The relation between the modulus and the quasi-periodicity is unknown to me, but the quasi-periodicity is not observed to be constant over long realisations of the series. Then, the Tribonacci series is simply a case of $m = 3$. The relations [6.2], [6.3], [6.4] will hold for subseries, that is segments between the operation of $mod(k)$, but not across that operation of [6.7].

The other modification of [6.1] is to write, more simply,

$$U_n = q^{-1}[U_{n-1} + U_{n-2} + U_{n-3}], \quad 0 \le q \le 1 \qquad [6.8]$$

which if $q \simeq 3$ becomes an AR(3) series, and if $q \ge 3$ it runs onto a point attractor. This one we will call **Tribarq**. It is less interesting, so it is eschewed here.

From runs 200 iterations long, with $U_1 = U_2 = U_3 = 1.0$ it is possible to estimate some statistical parameters of the Tribmodk process as a function of the modulus k. The estimates of the fractal dimensionality D and the Hurst exponent H are from an algorithm of West, Hamilton and West (2000) and their confidence limits are not known. MaxP is an estimate of the dominant periodicity from inspection of the autocorrelation spectrum; this is not necessarily constant over extrapolation. The largest Lyapunov exponent Λ_1 is from SANTIS. The entropic Schwarzian analogue ESf is as previously defined (Gregson, 2000), with partitioning constant .015. Slope refers to $mean\Delta^1 U$.

The D and H values are not independent parameters, because they are linked by the relation $D + H = 2$, and are included for illustration, but the

---

[2] The use of modular algebra in maps with recurrent trajectories is not, of course, new. It was used in Arnold's cat map in various forms to induce mapping on a torus (see Schuster, 1984, p. 152 et seq.)

Table 6.1
Estimates of some statistical properties of Tribmodk

Property	mod 7	mod 10	mod 13	mod 19
Mean	2.6061	3.3798	4.3795	6.2192
s.d.	1.9040	2.8795	3.9187	6.0812
slope	-.0002	+.0012	+.0037	+.0031
MaxP	40	19	9	41
D	1.8287	1.6925	1.6963	1.6562
H	.1713	.3075	.3037	.3438
Λ_1	+.1489	+.1305	+.0563	+.0519
ESf	-1.9495	-2.4398	-.6758	+.4396

remaining parameters each describe different properties of the series. The approach advocated here is to compute a diversity of measures to characterise adequately the peculiarities of the dynamics, including any evidence of nonstationarity. This necessitates the further exploration in Table 6.2.

Some of the estimates of properties in Table 6.1 are reminiscent of chaotic dynamics, so sensitivity to initial conditions is an aspect of Tribmodk that is worth examination. Perhaps the simplest way to investigate the effects of varying initial conditions is to make a second-order change in one of the original starting values, replacing U_1 by 1.01. In Table 6.2, this modified mod10 series is coded as S_1. Alternatively, we may take any three successive U_n values from its evolving trajectory as a new starting point; as the series is deterministic, this will replicate the gradual changes in the local quasi-periodicity of the trajectory if these new subseries are concatenated.

For example, the series S_2 is created from the mod 10 column in Table 1 by putting

$$U_1(S_2) \quad = U_{170} = 3.991 \quad [6.9.1]$$
$$U_2(S_2) \quad = U_{171} = 6.539 \quad [6.9.2]$$
$$U_3(S_2) \quad = U_{172} = 1.163 \quad [6.9.3]$$

Figure 6.1: Graphs of four Tribmodk series for contrast
Tribmodk with k= 7, 10, 13 and 19

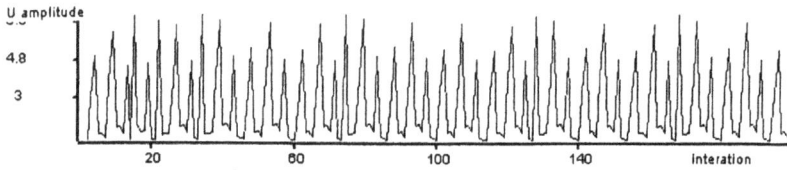

Tribmodk 07, 1-200

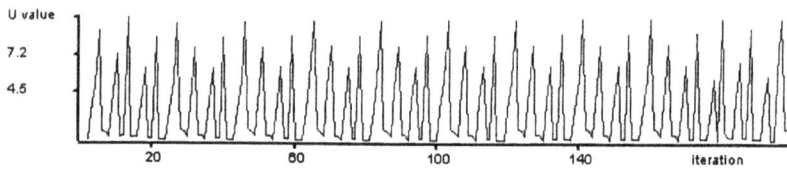

Tribmodk 10, 1-200

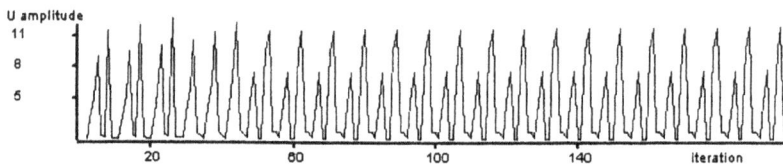

Tribmodk 13, 1-200

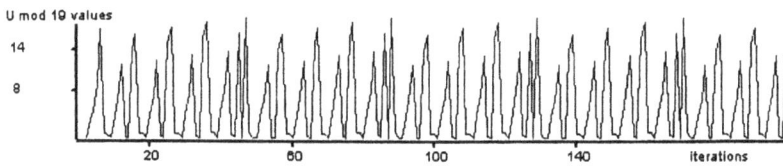

Tribmodk 19, 1-200

The values for S_3 and S_4 are similarly lagged, $U_1(S_3) = U_{340}$, $U_1(S_4) = U_{510}$, each series is 200 iterations long, so that they overlap.

Table 6.2
Variations in properties of Tribmod10 with starting values

Property	S_1	S_2	S_3	S_4
Mean	3.3874	3.3931	3.3946	3.3435
s.d.	2.8859	2.9068	2.8953	2.8139
slope	+.0012	+.0015	-.0005	+.0005
MaxP	19	9	9	19
D	1.6925	1.7012	1.8287	1.6882
H	.3075	.2988	.1713	.3118
Λ_1	+.1276	+.1120	+.1433	-.0010
ESf	-2.4398	-2.4128	-2.4128	-2.4181

Higher-Order Derived Series

Given any series x_j, $j = 1, ..., n$, it is possible to create other series by internal multiplication across lags. The simplest series is then

$$X_k := x_j x_{(j-k)}, \ j = 1, ..., n, \ k \text{ fixed} \qquad [6.10]$$

and the family of X_k with k a variable defines the bases of the autocorrelation spectrum of x.

If the multiplication is taken to a higher order, then

$$X_{(m,n)} := x_j x_{(j-m)} x_{(j-n)}, \ \forall m, n \ m \le n \qquad [6.11]$$

and $X_{(m,n)}$ is sometimes, in frequency domain analyses, called a bispectral series. If $X_{(m,n)}$ is computed over a set of m, n values, it gives us a triangular matrix of series, which, because the process [6.11] is symmetric in m, n, can be treated as the off-diagonal cells of a square skew symmetric matrix, $\mathcal{M}$ of size $n \times n$.

Each cell in the $X_{(m,n)}$ matrix may be replaced by a function of the series. In this context, we may compute the ESf of each of the $X_{(m,n)}$, and for convenience write

$$b(m, n) := ESf(X_{(m,n)}) \qquad\qquad [6.12]$$

The triangular matrix computed here for each of the Tribmodk series is of the form

$$
\begin{array}{lllll}
b(1,2) & b(1,3) & b(1,4) & b(1,5) & b(1,6) \\
b(2,3) & b(2,4) & b(2,5) & b(2,6) \\
b(3,4) & b(3,5) & b(3,6) \\
b(4,5) & b(4,6) \\
b(5,6)
\end{array}
$$

If this matrix is reflected to give the full square form, then its eigenvalues are all real (Gregson, 2001), because all the elements $b(m, n)$ are real and not complex. It should be noted that the values of m, n are minimal and thus the matrix acts like a high-pass filter. If n were to be set at 2, 7, 10 ,13, 19 respectively, then different values related to slower forced frequencies would be found.

Table 6.3
Eigenvalues for the Higher-order ESf

Eigenvalue	k = 7	k = 10	k = 13	k = 19
(1)	-1.7448	-1.9506	-1.4773	-1.6390
(2)	-.7661	-.4573	+.4536	+.6265
(3)	+.5663	+.4114	-.3589	-.4952
(4)	+.3580	+.2667	-.2449	-.1892
(5)	-.2283	-.1775	+.2440	+.1627
(6)	+.1290	+.0286	+.0016	-.0557

Tables 6.3 and 6.4 were computed using only the first 200 iterations of Tribmodk. It is seen in Table 6.3 that the second through fifth eigenval-

ues change sign as k increases. There is a reflection in the pattern of signs between $k = 10$ and $k = 13$.

Table 6.4
Higher-order ESf matrices for the Tribmodk series
Values in italics lie within the 95% confidence intervals for random surrogates

For k = 7

-.2534	*-.2598*	-.2823	-.0627	-.0244
-.3170	-.6473	*-.4080*	*+.3047*	
-.4117	*-.4135*	-.5552		
-.2605	-.0313			
-.2428				n: 4

For k = 10

+.0281	-.3663	-.3118	*-.1989*	-.2875
-.4510	-.5262	-.5393	*-.1329*	
-.3087	*-.4255*	-.1308		
-.5067	-.1954			
-.3430				n: 4

For k = 13

-.3703	-.1928	-.3817	-.1260	-.0824
-.1799	*-.3469*	-.3520	-.0276	
+.0662	*-.1696*	-.1210		
-.5422	*-.3138*			
-.3143				n: 4

For k = 19

-.3645	-.1545	+.0107	*-.2939*	-.4697
-.3226	-.3355	-.0869	-.3339	
-.0167	-.1036	-.5846		
-.4247	-.3677			
-.1273				n: 2

To show the pattern of the higher-order ESf values that generate the eigenvalues for the system, the triangular $b(m, n)$ matrices are set out in Table 6.4. The **n:** values in Table 6.4 are the number of italicised values out of 15 in that triangular matrix. There is some agreement between the ESf values in Table 6.1 and these **n:** values in that the $k = 19$ data are more different from the other three series; as k increases resemblance to a simple strictly periodic or random series diminishes. The emergence of a more complex wave form is also shown in the graphs.

Discussion

If a series is generated by a nonlinear process, then the set of statistics to characterize it adequately has to be more than the sufficient statistics of a linear process with Gaussian residuals. This is so even under stationarity of the parameters generating the observed realization of the trajectory. The simplest statistics, then, can only serve as an exploratory way of finding signs of suspected non-stationarity by examining successive sub-series. If there is no reason to expect a simple analytic representation of the under-lying processes generating the trajectories then confidence limits on any of the statistical indices, such as used here, can only be estimated by Monte Carlo methods.

If series are short, as is deliberately the case here, and have many singu-larities, then attempting a representation through Fourier analysis is very cumbersome and, even with wavelets, would be potentially misleading[3] Wavelets can handle Dirac δ functions when they arise within a reasonably long process, but the singularities induced here with the mod jumps are of a different form.

In Tables 6.1 and 6.2, the information provided by the slope merely indicates that the process has no detectable shift in its mean over the sampled series length. In that limited sense, it is stationary and might be within the trajectory of a closed attractor. The ESf uses none of the in-

---

[3] Kaiser writes, p. 61, of "the tremendous inefficiency of ordinary Fourier analysis in deal-ing with *local* behaviour" (Kaiser, 1994).

formation in the mean and s.d. and is sensitive to differences which are not sufficiently captured in the s.d. H has low values, which is usually interpreted to mean that the process is antipersistent (Feder, 1988).

The difference in terminal values for S_1 in Table 6.2 and $mod10$ in Table 6.1 is simply due to setting $U_1 = 1.01$ as against $U_1 = 1.0$. These values are respectively

$$
\begin{aligned}
U_{198}(S_1) &= 9.915, \text{ c.f. } 9.893 &[6.10.1]\\
U_{199}(S_1) &= 1.879, \text{ c.f. } 1.875 &[6.10.2]\\
U_{200}(S_1) &= 1.964, \text{ c.f. } 1.960 &[6.10.3]
\end{aligned}
$$

which indicates that trajectories initially separated do not converge onto the same values over the range explored. Nor, however, do they widely diverge, which is compatible with the small positive Λ_1 values.

The variations in Λ_1 and in ESf over successive epochs are small and may be due to computational approximations in the relevant algorithms, but there is evidence from recurrence plots that some period of non-stationarity does emerge within the longer series $U_1, ..., U_{510}$, and the pattern of peaks in the autocorrelation spectra changes and becomes more complicated with evolution. For the $mod19$ series, the autocorrelation spectra and the recurrence plot become even more complicated and qualitatively different from the lower $modk$ series. This is reflected more obviously in the ESf value, which becomes positive (this shift in sign has been found previously in varying parameters in edge-of-chaos series), as compared with the changes in the other statistics.

It is obvious from Table 6.1 that ESf is very sensitive to the choice of k, but equally within one k value it may be insensitive to local second-order changes in the evolution of the same trajectory, as seen from the twin values of -2.4128 in Table 6.2. This insensitivity can arise as a consequence of the choice of the partitioning constant δ (See Appendix). The higher-order ESf matrices can be used to compare the dynamics in successive short concatenated subseries, particularly to explore for evidence of nonstationarity, as they can reveal subtle changes that are missed by the lower-order parameters; this has been done for some EEG series (Gregson, 2001).

The conventional way of comparing the series, as they are obviously quasi-periodic, is to compute their Fourier Power Spectra, provided that we bear in mind the warnings just given. For illustration, spectra for Tribmodk $k = 7$ and 19, iterations 1-200, are shown; the extra secondary peaks that emerge as k increases are apparent, as is a shift in the location and power of a major peak in the middle range.

Figure 6.2: Graphs of Fourier Power Spectra
for Tribmodk k = 7 and 19

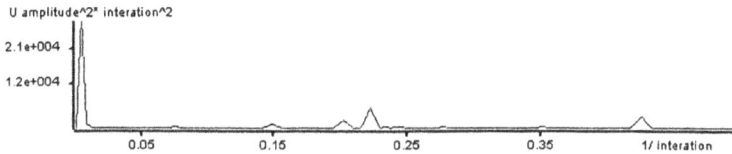

Tribmodk 07, Fourier Power Spectrum

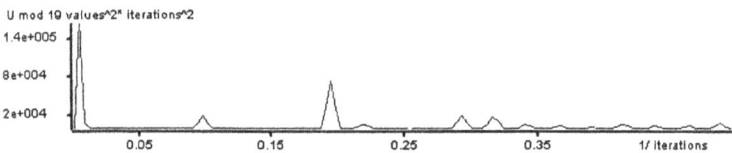

Tribmodk 19, Fourier Power Spectrum

The computation of D, H and Λ_1 over very short series is associated with difficulties, even if only relative values are sought. There is, however, some evidence that in such circumstances, with series appreciably shorter than the 200 used here for illustration, that ESf can still be informative, and can differentiate between stochastic and some nonlinear deterministic trajectories (Gregson, 2000b). It is trivially simple, using ESf, to differentiate between strictly periodic series and quasi-periodic ones, because for strictly periodic series the successive differences eventually vanish. They do not vanish for Tribmodk, which can have a somewhat curious distribution of $\Delta^4(U_n)$. Tribmodk is an example of a process which is strictly stationary in its parameters but, if examined by calculating the range of descriptive statistics now available for characterising non-linear dynamics, will appear in the light of some to be non-stationary in its outcomes. Such processes are not novel, the SETAR series created by Tong (1983) are cases in point.

The reader might appreciate some hint as to how one might use the Tribmodk family and its associated statistics m, k in experimental psychology. Perhaps the simplest way is to use a series extrapolation task, where the theory is used to generate a stimulus display over time and then the subject is asked to continue the series for another large number of trials[4]. This has been done by various authors with the logistic or Hénon functions (Neuringer & Voss, 1993; Smithson, 1997; Ward & West, 1998; Heath, 2002), and comparisons made between the recurrence plots of theoretical and response time series. Recurrence plots are not helpful for short series, because of their sparsity, but other statistics such as ESf can be used for comparisons of theoretical and response series, even when the series are replete with singularities.

There are deep relations between the chaotic dynamics of complex cubic maps on the unit interval, their attractors and the Fibonacci series, and the Schwarzian derivative, that have been explored by Milnor and his coworkers (Lyubich, 1993). This suggests that, in terms of topology, there are a number of unanswered questions about the series used here. It is an open question whether Tribmodk should be called chaotic, in the strict sense used to describe a strange attractor, despite the positive value of its estimated largest Lyapunov exponent. It is a closed trajectory on an invariant manifold and, with increases in the one parameter k, it becomes more complicated, but not in the period-doubling sequence of, say, the Ruelle-Takens series. It is, perhaps, preferable to call it simply nonlinear deterministic and stable. Other such series with local memory have been also described (Morariu, Coza, Chis, Isvoran & Morariu, 2001) in com-

---

[4] This type of experiment has also found use in clinical diagnosis (Helmuth, Mayr & Daum, 2000) with some simplification.

paring and contrasting human with theoretical memory processes. There are yet other ways of creating chaotic sequences by considering quadratic maps with no real zeroes. The main result is that rational initial conditions produce finite or infinite periodic sequences, whereas irrational ones yield infinite but not periodic sequences (Carvalho, 2002, p. 31).

Classical ARMA analyses

It can be informative to see what linear ARMA(p.q) representations of the Tribonacci series look like. The solutions are limited to models with $p \not> 5$, $q \not> 5$. We note that the order of fitted q increases with k, but the values of the p terms change irregularly.

Table 6.5: ARMA parameters for the four Tribonacci series

parameter	Trib07	Trib10	Trib13	Trib19
AR1	-.2203	.3214	.1125	.1049
AR2	.2169	-.3448	-.2658	.0674
AR3	.1025	.6165	.1415	-.1006
AR4	.2961	-.1432	.4127	.0387
AR5	.6046	.5498	.5986	.8857
MA1	-.0067	-.5429	.2484	-.1373
MA2	-.6222	-.1241	.0752	-.1031
MA3	-.2929	-.5964	-.1545	.1808
MA4		.3003	-.2238	.2368
MA5			-.8825	-.9437
-2logΛ	.7181	.9168	.9675	1.1976

All the -2logΛ terms in both Tables 6.5 and 6.6 are to be read as × E+03.

The point of this analysis is to show that, if we have a purely deterministic process, which is, if we know its generating function, perfectly predictable, and we instead assume falsely that it is purely stochastic and stationary, then an ARMA model is computationally derivable. But this model tells us nothing about exactly how the process evolves; it is descriptive and not functional. It discriminates between the forms of Tribmodk if k is increased, but does so in a way that is much too complicated compared with the generating function. It thus violates Occam's Razor.

Paradoxically, this ARMA analysis could have some use, though it furnishes no real insights. If Tribmodk is used as the seed process in a serial extrapolation experiment and the subject is in fact capable of extrapolating within some tolerance bound, ϵ, then the ARMA model of the subject's extrapolation should resemble an appropriate minimum order ARMA(p,d,q), and that should also be a good fit to the ARMA(p,k,q) model of the generating seed series. We note that it is highly implausible that the strategy used by a human observer would be isomorphic with an ARMA model with 8 to 10 terms in a linear expression, as shown in Table 5. The brain does not store information that way.

Differencing at lag k, as shown in Table 6.6, does not make much change to the overall goodness of fit index -2logΛ, but the structure of the ARMA models does change, and the differences between Tribmod13 and Tribmod19 are apparent, particularly the collapse of the order of the MA terms. It is worth re-emphasising that these induced complicated patterns are all the consequence of changing just one real positive scalar parameter in Tribmodk.

ApEn modelling

Pincus, in a series of analyses (see, for example, Pincus, 1991, Pincus and Singer, 1998), has shown that a measure of approximate entropy (ApEn), which can be applied to relatively shorter time series, will qualitatively distinguish between some stationary processes, where other measures such as LLE or D2 are not reliably computable. It can be shown from a variety of cases that ApEn is not monotonic on ESf, and that ESf can distinguish between some pairs of series, including the Tribmodk series used

Table 6.6: ARMA parameters with differencing at level k

Parameter	Trib07	Trib10	Trib13	Trib19
AR1	.1283	.4038	.5632	-.5944
AR2	-.0799	-.2599	-.4054	-1.4286
AR3	-.6251	-.0374	.3901	-.9088
AR4	.5806	-.7429	-.8611	-.7501
AR5		.6015		-.6007
MA1	-.2608	-.6675	-1.0313	.3037
MA2	-.7149	-.1765	-.2128	.9979
MA3	.6554	.0764	-.0215	
MA4	-.2574	.6673	1.2268	
MA5	-.4073	-.8079	-.7046	
$-2\log\Lambda$	.7395	.9384	.8470	1.1267

here, rather better than ApEn. The two measures do not use the same data properties.

The estimated ApEn values, using $m = 2, r = .20$ (the parameters of choice in ApEn), for the Tribmodk series are, in order of increasing k,

$$.5133, \ .5625, \ .5143, \ .5117$$

These values should be compared with the ESf values in the last row of Table 6.1.

Modified Tribonacci series with one control parameter k are strictly deterministic. If k is replaced only when it serves as a divisor by a random variable θ, where $1 \leq \theta \leq k$, then the process becomes different in its evolution, but is still bounded and edge-of-chaos. Numerical comparisons of solutions to the ESf and ARMA modelling of the series are provided; the effects of introducing stochastic noise appear to be irregular and to be diminished as k is increased.

There have been a number of studies concerned with the identification and analysis of processes that are reasonably believed to be a mixture of nonlinear dynamics and random noise. Such processes may be treated as a mix of low dimensionality dynamics and high dimensionality stochastic background, or as nonlinear dynamics masked by stochastic noise of indeterminate structure. The degree of masking is critical for identification of the deterministic components. In short data samples, it may be impossible to disentangle signal from noise if we regard the nonlinear dynamics as generating the true signal. Misidentification of dynamics thus arises if one uses methods that presuppose the presence of, and only of, stochastic dynamics. At the same time, biological processes are considered to be contaminated ubiquitously with noise, or even require for their viable plasticity the presence of random components (Schaffer, Ellner & Kot, 1986).

The Tribmodk series could be interpreted as a model of some cognitive process. In fact, if it is used as a stimulus generator for serial extrapolation tasks, then, if the subject does extrapolate most of the properties of the series whose local evolution has been presented, it can be thought of as evidence that the brain can perform the operations intrinsic to Tribmodk.

Tribmodk is built on three deterministic operations: (i) storage of a few numerical values in short-term memory, (ii) addition, and (iii) rescaling by something like shifting a decimal point or division by an integer. These are all simple operations that are used in AI neural networks and thought to be performed in real neurophysiological connected structures (Oprisan & Canavier, 2001). It is important to note, from inspection of the graphs in the previous study, that Tribmodk depends on its starting values for a few terms and then runs onto an attractor.

The parameter k in fact performs two distinct roles: a time marker, which has been called a modulus but which triggers the action of the divisor, and the divisor itself. Let us relabel these k_t and k_d respectively, in effect increasing the number of free parameters in the model by one. Now let us replace k_d by θ_d, where $1 \leq \theta_d \leq k$ and $\theta_d \sim rect((k+1)/2, .289(k-1))$, where $rect$ is a random i.i.d. distribution on the interval. In doing this, a mixed deterministic-stochastic process has been created. As each of k_t and k_d could be replaced by a bounded random variable, there are four possible Tribmod models; deterministic $k_t k_d$, mixed $k_t \theta_d$, $\theta_t k_d$, and stochastic $\theta_t \theta_d$. The purpose of this part is to compare the properties of Tribmodk and Trbmod$k_t \theta_d$.

A secondary modification is to admit the range of θ not over $1 \leq \theta \leq k$ but only over $k \pm 1$, which is illustrated in some of the figures. Obviously, this is less disruptive of the dynamics.

ESf Analyses

The ESf analyses are computed in the same was as for the deterministic series; there is a complication in that any realization of the series now is not an exact replication of any other with the same seeds x_1, x_2, x_3 and k, so there is variance in the evolution of the source trajectory as well as in the surrogate randomizations.

In Table 6.7 it appears that, for $k = 13$, the process is more consistent with being random. In all the other three series it is nearer to chaotic than random, (as ESf is more positive, as shown in Gregson 2002) but the difference (-.539 versus -.549) for $k = 19$ is very slight.

Table 6.7: ESf and surrogate c.i. for the four Tribmodkθ series

surrogate: bounds: k	ESf	95% c.i. lower	95% c.i. upper
7	-.2832	-.450	-.418
10	-.4801	-.616	-.588
13	-.6736	-.675	-.619
19	-.5390	-.583	-.549

$\delta = .03$ in all cases.

The interplay of k and θ on the emergent dynamics is obviously not simple.

The higher-order ESf analyses may also be compared with the previous results and are tabulated in the same fashion.

The magnitudes and signs of the Eigenvalues are changed and relatively diminished. There is a suggestion that the patterns converge as k increases and the effect of the presence of a random component is lessened.

Table 6.8: Eigenvalues for ESf[b(m,n)]

Eigenvalue	k=7	k=10	k=13	k=19
(1)	-.6165	-1.4637	-1.1527	-1.1440
(2)	-.4709	+.2256	+.4969	+.3358
(3)	+.2772	-.2219	-.3054	-.2443
(4)	+.1827	+.1590	-.2035	+.1009
(5)	+.0952	-.1124	+.0706	-.0954
(6)	+.0078	-.0368	-.0243	-.0420

It can be seen by comparing **n:** values in Table 6.4 of the Tribmodk analysis with Table 6.9 that the effect of introducing stochastic variation θ_d is not to increase the higher-order random structure but, as k_t increases, slightly to diminish its randomness.

Classical ARMA analyses

The tables in this part of the note should be compared with the corresponding tables 6.5 and 6.6 previously considered for long memory only, for the four k values and for undifferenced and differenced ARMA models. The same constraint is used, that only models of ARMA(p,q) with $p \not> 5, q \not> 5$ are computed to find a best fit within those limits.

Table 6.9: Triangular ESf[b(m,n)] matrices for Trinmodkθ
Italicised values lie within the random surrogate 95% c.i. bounds

For k=7

-.1310	-.1994	*-.1358*	.0445	*-.1109*
.0367	-.1878	-.2196	-.0910	
-.1527	.2472	*-.1457*		
.0405				**n: 4**

For k=10

-.3367	*-.1850*	-.3339	*-.1348*	-.2787
-.2045	-.2485	-.2570	-.1996	
-.4252	-.2512	-.1097		
-.2282	*-.2186*			
-.2134				**n: 5**

For k=13

.0031	-.2150	*1953*	-.3185	-.3603
-.3413	-.2327	-.2872	-.2872	-.3015
.0804	*-.1082*	-.0822		
-.1800	-.0863			
-.1711				**n: 2**

For k=19

-.2785	-.2097	-.2237	-.1317	-.1390
-.1622	-.2608	-.1640	.0199	
-.5257	*-.1423*	-.0928		
-.1280	-.2975			
-.0866				**n: 2**

Table 6.10: ARMA parameters for the four Tribonacci series

parameter	Trbrn07	Trbrn10	Trbrn13	Trbrn19
AR1	.6992	.4719	.4359	-.2965
AR2		-.9970	-.5315	-.4208
AR3			.5307	-.4535
AR4			-.6697	-.3889
AR5			.2052	.4549
MA1	-.5126	-.8233	-.5541	.0375
MA2	.1195	.7459	-.3750	.0039
MA3	.3158	-.2045	-.6385	.0574
MA4	-.1849	-.4181	.9332	.1645
MA5	.2753			-.8433
-2logΛ	1.0339	.8869	.9042	1.1638
LLE	+.0915	+.1703	+.1660	+.0980

All the -2logΛ terms in both Tables 6.10 and 6.11 are to be read as $\times$ E+03.

When the differenced series are compared, k_d fixed versus $k_t\theta_d$, it is found that, as k increases to 19, the differences in the ARMA estimated structure become relatively quite small. It appears that the deterministic part of the process begins to dominate the effects of randomization of the divisor θ. The differenced series are curious in their form and can be more irregular and less apparently periodic than the undifferenced series.

The remainder of these Tribmodk and Tribmodkθ comparison Figures 6.11 through 6.25, concerned with recurrence plots, and with some delay plots are given at the end of the chapter. Note that in Figures 6.24 and 6.25 another form of Tribmodkθ has been added in which the noise in the divisor θ is kept in the narrow range $k \pm 1$. The variance of the noise distribution is a critical parameter in determining if the nonlinear dynamics are preserved or masked. Obviously a range of cases between this narrow rectangular p.d.f. and the wide one $(1, ..., k)$ can be created and explored. If the task is to discriminate between series, each already known to be of

Table 6.11: ARMA parameters with differencing at level k

parameter	Trbrn07	Trbrn10	Trbrn13	Trbrn19
AR1	.1738	1.7685	-.2762	-.5641
AR2	-.0688	-.9417	-.0947	-1.3903
AR3	-.5756		.1333	-.8412
AR4	.5559		-.5687	-.7107
AR5			-.6851	-.5675
MA1	-.3206	-2.0710	.0448	.2976
MA2	-.7266	1.5726	-.6052	.9995
MA3	.6417	-.3288	-.6367	
MA4	-.0946		.7369	
MA5	-.4999		.5236	
-2logΛ	.7399	1.0603	.8688	1.1270

The following series of graphs summarise pictorially the qualitative features which vary with the choice of the parameter k. In these cases, k is a selective filter and, to match real data properties and exploration over the whole field of k and θ, solutions would be needed. The same selected values of k have been used, prime numbers were chosen to block factorization effects.

---

the Tribmodk family, with k as the variable, it may be regarded as as estimation problem and not an identification problem.

Discussion

As remarked, this part of the study arose out of reconsideration of experiments in which it was asked if subjects could serially extrapolate a chaotic trajectory. The answer appears to be that they could not strictly do that, but that they could do something which preserved some qualitative features, such as quasi-periodicity or anti-persistence, as quantified by the Hurst in-

Figure 6.3: Tribmodkθ 07 series, 1,200

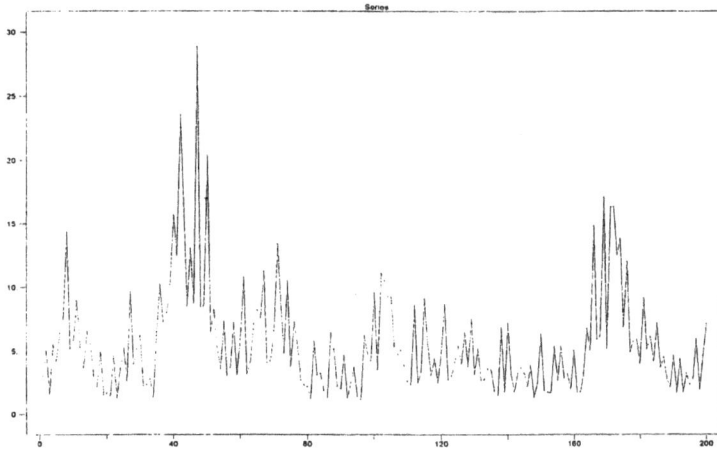

Figure 6.4: Tribmodkθ 07 differenced at 07 series

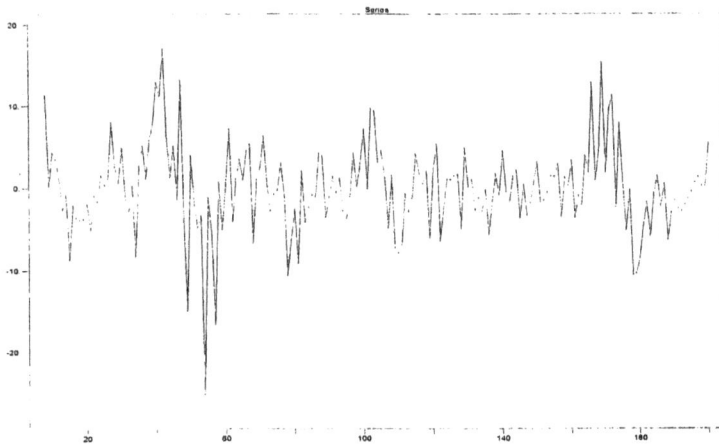

Figure 6.5: Tribmodkθ k=10 series, 1,200

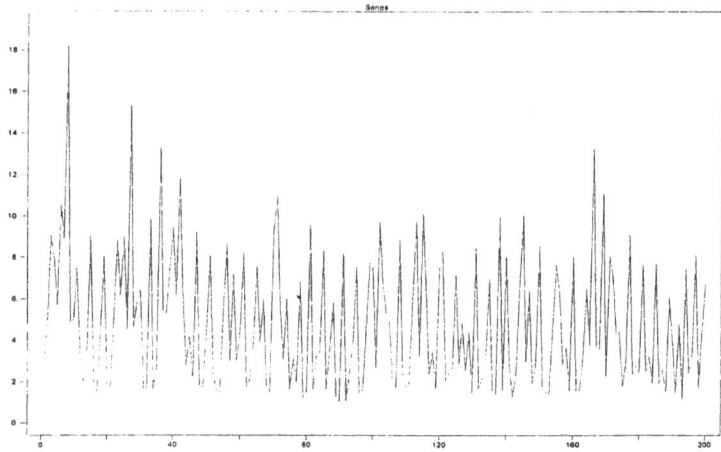

Figure 6.6: Tribmodkθ k=10 series, differenced at 10

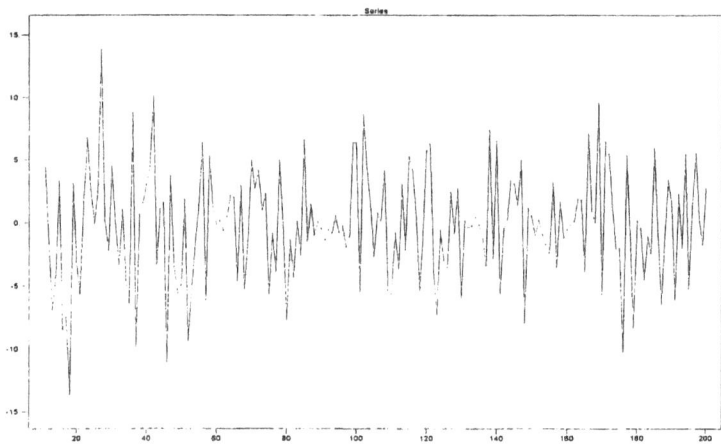

Figure 6.7: Tribmodkθ k=13 series, 1,200

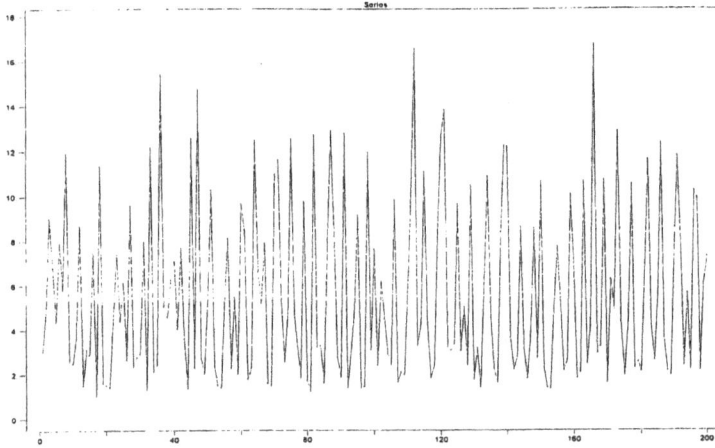

Figure 6.8: Tribmodkθ k=13 series, differenced at 13

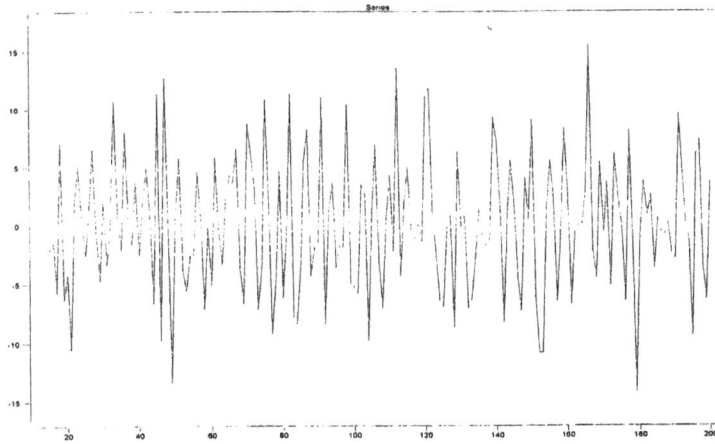

Figure 6.9: Tribmodkθ k=19 series, 1,200

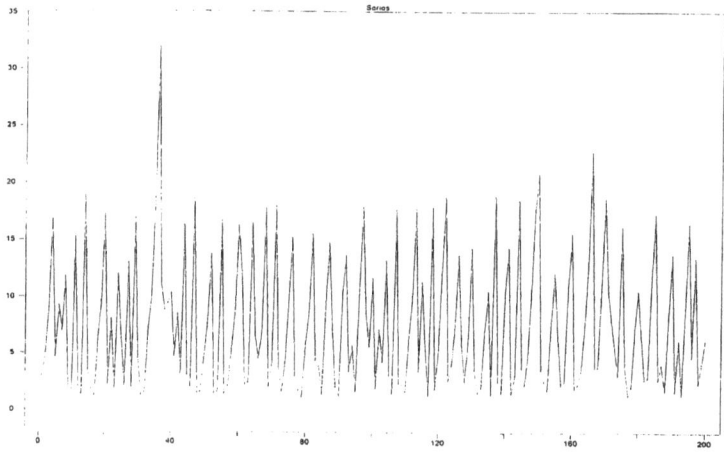

Figure 6.10: Tribmodkθ k=19 series, differenced at 19

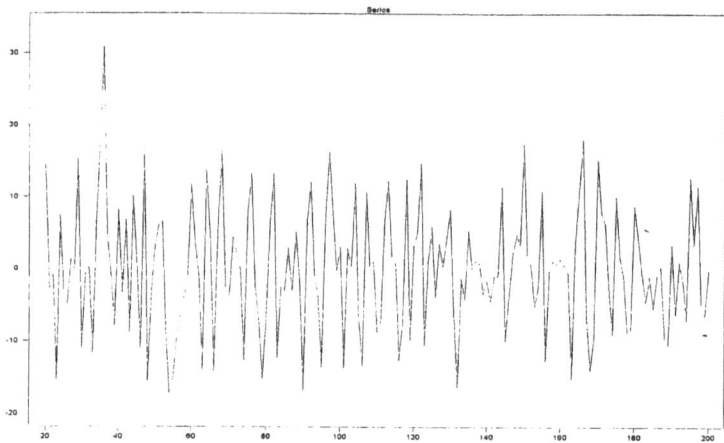

dex. It is easier to extrapolate locally from a chaotic series generated from the logistic map than from a random series.

The Tribmodkθ series created here might be thought of as a nonlineary dynamic repeatedly disrupted by transients; in this sense it has some ecological validity. That argument has been appealed to by Gregson (1995) in the context of nonlinear psychophysics but using the Γ recursion. The point being made here is that using the logistic attractor after reaching stability on the invariant manifold is an extreme form of stability. Even so, trajectories from stationary nonlinear dynamics can have subseries which are qualitatively identifiably different from the rest of the series (Warren, Bashford, Colley & Brubaker, 2001); if these are what are presented as the stimulus series the results could hardly be the same as from some other subseries from the same trajectory, though the subseries themselves may be identifiable by the observer.

Four studies that have focussed on the serial extrapolation problem are by Neuringer and Voss (1993), Metzger (1994), Smithson (1997) and Ward and West (1998). These all used the logistic function with the gain constant set near 4 to produce a trajectory that was stable. Heath (2002), however, used a Hénon attractor to generate a discretized trajectory. The details about what subjects were asked to do, in terms of length of stimulus series, length of extrapolation, comparison with random series, and feedback provided or withheld, differ between studies. Heath importantly showed that some subjects could discriminate (in terms of very local serial extrapolation) a chaotic series from one with the same spectral density that was not chaotic. The mode of response and the precision of responses required also varied. The analysis of results was in some cases presented as the lag-one plot $x_i \mapsto x_{(i+1)}$, which takes a very simple form for the logistic function. Metzger showed that if precision of response is relaxed then a simple rule, that can be learnt from inspection of the stimulus series, will approximate the lag-one delay plot. No such simple rule does generally exist, and in some of the Tribmodk and Tribmodkθ examples used here it is not in such a tidy form. In short, the logistic function is not a valid basis for making assertions about chaotic trajectories in general.

But Metzger's demonstration raises another important point; she showed that the delay plot is roughly approximated if the observer uses a small, and thus finite, set of rules of the form "if j occurs then follow it by k". This gives a series of lumps which lie roughly on the continuous curve of the logistic delay plot, a parabola for the gain parameter value selected. But such a set of rules are the cells of a transition probability matrix; if that matrix is ergodic and complete, and the mapping is from t to $t + 1$, so no diagonal cells are absorbing states, then it may generate quasi-periodic cycles. In short, it resembles a Markov process and it may be anti-persistent. It is thus not incompatible with Smithson's later analysis. The question is, as originally put in these series extrapolation tasks, can the extrapolation from a chaotic trajectory be itself chaotic? The answer, if subjects are doing what Metzger identified, is no, because (Martelli, 1992, p. 206) "the conditions which qualify a system as chaotic require the presence of aperiodic orbits, but are not met by any dynamical system governed by a $q \times q$ matrix M." To explore this point further here we need Markovian approximations to the Tribmodkθ series, for an arbitrary number of states q. These and the terminal state vectors at ∞ are shown in the table, all assuming an equiprobable starting (rectangular) vector. The assumption is that five states are sufficient to represent the upper limit on memory storage that a subject would use to construct rough transition rules. The patterns suggest that Tribmod19θ would defeat a subject attempting extrapolation on that basis.

Human Predictive Capacities

As noted, a number of studies have addressed the question as to whether the human subject can serially extrapolate a series that is either random or chaotic and not, therefore, strictly periodic or stationary. We know that observers have difficulty in recognizing what is a random sequence and attribute to local features evidence of non-randomness that is not statistically valid. But recognition and production of series require different abilities that are not tightly coupled: I can tell often that music is by Mozart and not by Brahms, but I can play neither.

Table 6.12: Markovian representation with 5 states
Transition probability matrices for Tribmodkθ series

d1	d2	d3	d4	d5	∞
	Tribmod07θ				
.82	.15	.02	.01	.00	.736
.59	.28	.10	.00	.03	.190
.33	.25	.33	.00	.08	.059
.00	1.00	.00	.00	.00	.005
.00	.50	.50	.00	.00	.010
	Tribmod10θ				
.55	.24	.21	.01	.00	.558
.54	.24	.14	.04	.04	.246
.62	.29	.09	.00	.00	.171
.67	.33	.00	.00	.00	.015
.50	.00	.50	.00	.00	.010
	Tribmod13θ				
.47	.15	.20	.15	.04	.488
.46	.22	.12	.17	.02	.201
.58	.29	.06	.06	.00	.156
.46	.31	.15	.08	.00	.131
.60	.00	.20	.20	.00	.024
	Tribmod19θ				
.65	.00	.00	.15	.20	.563
.76	.24	.00	.00	.00	.143
.00	1.00	.00	.00	.00	.035
.32	.60	.08	.00	.00	.122
.37	.00	.19	.26	.19	.137

There is a little-known historical precursor to such experiments, in which Whitfield (1950)[5] asked subjects to respond to an imaginary questionnaire, to which only their responses were recorded; the questions were private inventions of the subject and not revealed. He found that the statistical structure of their response sequences, made in yes-no or rating scale fashion, changed with the length of the questionnaire. In modern parlance, they moved onto an attractor over time. The important thing about this neglected study is that there was nowhere any mention of randomness, but the subjects were free to run onto an attractor of their own devising and did so. The results showed that the series were autocorrelated, to some degree, and unbalanced in terms of the proportion of 'yes' and 'no' responses to an imaginary questionnaire which only allowed for binary responses. There were also warm-up effects. These features match what is reported in random series production experiments; it suggests that calling a series random may be irrelevant, but that the human respondent does involuntarily converge onto a noisy attractor. Call that attractor Q, and its trajectories on its manifold the set $\{t(Q)\}$.

Serial extrapolation without feedback has some common features with the imaginary questionnaire, what subjects do after they have only their memory to rely on for extrapolating a quasi-periodic series, (largest Lyapunov exponent positive and exponentially decaying local predictability) is to move to an attractor with intrinsic instability. Even for extrapolating the simplest sinusoid, as in motor skills, there will be second-order departures from a limit cycle and response sequences that locally return to the deterministic trajectory, such as overshooting or delays. Such patterns have been studied from the standpoint of ODEs of oscillations (Hale, 1963; Hoppensteadt, 1979).

If the stimulus series is presented continuously in parallel with the response series, then it may take on the role of a forcing function, not necessarily at a conscious level. This requires a different method of analysis (Gregson and& Leahan, 2003).

Suppose that the original stimulus series is generated by a determinis-

---

[5] This work was done at University College, London, at the time I was a student.

tic attractor $\mathcal{A}$ and that the subject has in her own response repertoire $\mathcal{A}^*$ where $\mathcal{A}^*$ has two properties, it is contaminated with stochastic noise and it is in some sense near to $\mathcal{A}$. In fact, what we can only observe as subject, but not as experimenter, are trajectory samples, $\tau(\mathcal{A})$ and $\tau(\mathcal{A}^*)$.

Various questions then arise: if, instead of $\mathcal{A}$, we present $\mathcal{A}^*$, does this make the serial extrapolation task easier or harder? To put it another way, if the subject already has both $\mathcal{B}^*$ and $\mathcal{A}^*$, does the presence of $\mathcal{A}^*$ make it easier to avoid using a mismatch $\mathcal{B}^*$? What constitutes a match or mismatch for the task is in terms of trajectories $\tau(\mathcal{A}) \simeq \tau(\mathcal{B}^*)$, not direct matching, though some attempts have been made subsequently to match by Poincaré diagrams. It is assumed that a subject can not produce a purely deterministic nonlinear trajectory by making motor or verbal responses and it is doubtful if a linear process can be produced without some error tolerance.

The family Tribmodk with variable k produces trajectories that do not simply resemble one another; from comparing some samples the subject cannot readily see that only one parameter is involved in generating the diversity of patterns, re-exploration of a family of perceptually similar patterns is not necessarily an optimum search strategy to find a valid serial extrapolation sequence. If only a family $\{\mathcal{B}^*\}$ exists, then the subject seeks within that family to find a close match to the presented $\mathcal{A}^*$ by varying some parameters in a subset of $\{\mathcal{B}^*\}$. As soon as noise is admitted, the distinction is between $\exists \mathcal{A} \cap \mathcal{B}$ which may not be a condition that could be satisfied, and $\exists \mathcal{A}^* \cap \mathcal{B}^*$ but $\mathcal{A} \cap \mathcal{B} = \emptyset$ which can be satisfied rather trivially if there is no upper bound on the noise variance.

There is yet another way of conceptualising the conflict between two trajectories: as an invasion by the stimulus attractor of the attractors already dynamically resident within the observer. By an analogy with ecological models (Geritz, Gyllenberg, Jacobs & Parvinen, 2002), the stimulus series is a mutant invasion, which may be absorbed, continue in parallel, or overtake the system. Such replacement of the initial dynamics is implausible in the context of the serial extrapolation experiments, they are not like learning to play a Beethoven sonata that is long, complicated and remembered.

What subjects can actually do with simpler oscillating series as a tracking rather than an extrapolation task has been surveyed by Large and
Jones (1999). If the subject has alternative choices of series to employ, then
it becomes a sort of discrimination task for quasi-periodic series; Meinhardt (2002, p. 143) observes in an analogous context that "discrimination
learning can be understood as a higher level process of gradual refinement
of code selection out of a rich code base provided by lower level stages".
The problem is that no lower level stage may lead by revision to an attractor that is synonymous in some sense with a chaotic trajectory.

There have been numerous attempts in statistical and physical theory to identify the dynamics of nonstationary time series for example,
by Machens (2002) in physics and Fuentes (2002) in spatial processes but
these are useless for the short series of psychological data, though sugges-
tively they may employ information measures, or interesting convolutions
of stationary processes that become nonstationary in the long run.

In formal terms, the problem and some of the observable results for
Tribmodkθ, $\theta \sim rect[k \pm 1]$ here are related to the theory of stability under
periodic disturbances (Hoppensteadt, 2000, p.106); under some conditions
the trajectories of the perturbed system lie sufficiently close to those of the
unperturbed system. The examination of the recurrence and delay plots
gives some qualitative idea of what noisy disturbances are tolerable in the
stimulus series.

It is known, from analyses of the more standard nonlinear attractor dynamics (Logistic, Henon, etc) that a small introduction of stochastic noise
does not perturb the closure of limit cycles, but a larger noise level will
eventually destroy the attractor structure (Kapitaniak, 1990). A deterministic trajectory is replaced by a noisy orbit, which is what we have done
here with Tribmodk. But if the noise is slight, the largest Lyapunov exponent remains positive and its fluctuations are smoothed out. The separation of trajectories in chaotic dynamics (associated with the positive LLE)
is accompanied by noise which is well-known in diffusive processes. In
this context, compare the 2D delay plots from Tribmodk and Tribmodkθ
with $\theta \sim rect[k \pm 1]$, and with Tribmodk$\theta$ $1 \leq \theta \leq k$. The narrow band
noise appears to clarify the delay plot structure, the wide band noise de-

stroys it.

There are yet more difficulties, in that if we did have a long enough response series to evaluate its properties it could not with certainty be concluded that it was chaotic, even if the original stimulus series was itself chaotic and the two appeared to match. Hoppensteadt (op cit., p. 200) notes that some functions of a slow almost periodic form can be indistinguishable from chaotic series in practice, say chaotic series that have a power spectrum supported by an interval of frequencies. So comparing the stimulus series and the extrapolating response series by comparison of Fourier spectra may be an inappropriate test.

The psychological question that remains is what, if anything, in human performance resembles the mixed deterministic-stochastic evolutions synthetically generated here when cognitive tasks are repetitively performed? If serial extrapolation is induced from a mixed system and mixed systems are characteristic of human performance, then the problem is to measure the disparity (or match) in the dynamic structure of two mixed systems and to find measures of nearness of the two systems from relatively short time series. The approach illustrated here creates reference patterns from Tribmodk and Tribmod$k_t\theta_d$ and would then require that data from real human extrapolative behaviour was analysed in exactly the same way. There is no known single sufficient statistic to characterise the evolution of mixed nonstationary processes and no reason to think that one would exist, though attempts have been made to devise such a measure on a structural equation which has mixed dynamics (Kauffman & Sabelli, 1998).

Conclusions

In the serial extrapolation task there are a number of different mechanisms involved:

(1) The local trajectory from a known but — to the subject — unspecified attractor $\mathcal{A}$,

(2) Some attractor B^* (or one of a set of such attractors between which switching might spontaneously occur) in the subject that is perturbed from its manifold by transients from other attractors,

(3) The storage of A as A^* in short-term memory,

(4) a modification or replacement of B^* by A^* where these two share locally some coarse trajectory properties, as a sort of learning. It is this stage where something like a Markovian process may be substituted

(5) a drift back to the reinstatement of B^* as the resting state of the observer's system dynamics.

It is (4) that presents problems, for various empirical and mathematical reasons (Buescu, 1997; Wiggins, 1992; Hoppensteadt, 2000). The operation of one attractor on another can be induced by a forcing function, by a transition over a heteroclinic orbit, or by more exotic processes that are not demonstrable except in simulations with very long running times[6]. There is no way with short data series that resolution of these ambiguities can be solved. The few properties of the observed time series that have been used to match the stimulus and extrapolation response series are not specifically characteristic of nonlinear dynamics, and indices that might be of use, such as LLLE, are not computable on short series, indices such as ESf that are computable are not fixed scalars for most nonlinear evolutions, even under stationarity in the sense of being on an invariant manifold.

Note

The employment and meaning of the Schwarzian derivative (Schwarz, 1868) in nonlinear psychophysics was noted previously (Gregson, 1988, p. 27). It measures a property sometimes called "expansiveness". In one

---

[6] Concepts such as riddled basins or distinctions between different forms of attractor have found no use in attempts so far to apply chaos theory to psychological processes, as distinct from physiological or ecological modelling.

dimension, it is a natural measure of non-projectivity of a transformation f (Sullivan, 1983, p. 727).

There are fundamental differences between the information in a frequency analysis and that in the ESf representation[7]. The frequency domain representation, by Fourier's theorem, linearly partitions the series into parts of an infinite sum

$$f(x) := \sum_{i=1}^{\infty} F(\alpha_i, \pi_i, \theta_i) \qquad [6.11]$$

where α is amplitude, π is periodicity of a sinusoidal component F, and θ is phase. The series is truncated to sufficient terms to give an approximation when the data series is of finite length. The α information is removed by [6.1], the phase θ is not considered at all and the periodicity π is effectively partitioned into slow and fast components so that the information in the fast components dominates the construction of Tribmodk. The slow dynamics function as a carrier and, if they are sinusoidal, then their second and later absolute differences vanish and for the slow parts $ESf \rightarrow 0$. The representation of nonlinear trajectories as being a consequence of slow/fast psychological dynamics has been found informative. It is also noteworthy that approximate entropy measures, such as ESf, have found significant diagnostic use for dealing with local transients in EEG records in anaesthesiology (Bruhn, Ropcke, Rehberg, Bouillion & Hoeft, 2000).

---

[7] The shift from derivatives of a continuous function to discontinuous sequences is not innovative; see Chalice (2001) on the analogues of differentiation in discrete sequences and Sandefur (1990) for applications in dynamics

Figures 6.11 and 6.12: Recurrence plots for Tribmod 07 and Tribmod 07θ

The effect of θ randomization is to destroy the broken 7 periodicity. Note the initial unstable period before the process runs onto an attractor on the invariant manifold.

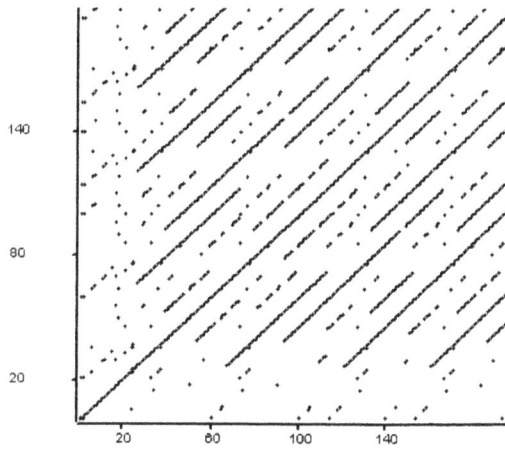

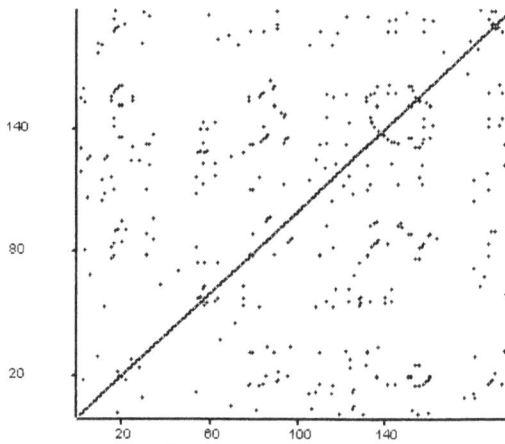

Figures 6.13 and 6.14: Recurrence plots for Tribmod 10 and Tribmod 10θ

There is a suggestion of non-stationarity here, but otherwise it is not interesting. Perhaps this is because here k is not a prime number.

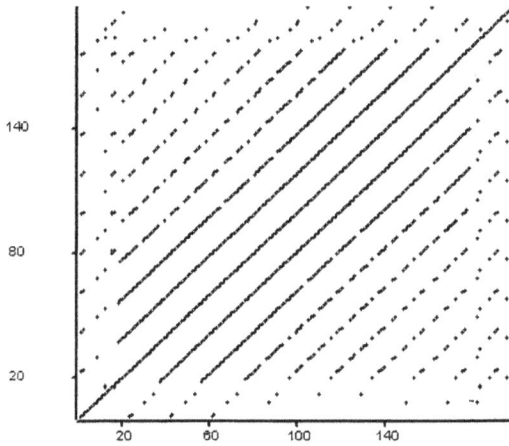

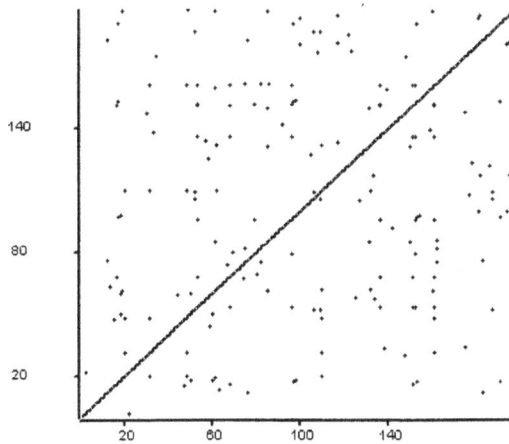

Figures 6.15 and 6.16: Recurrence plots for Tribmodk 13 and Tribmod 13θ

This is perhaps the most curious pattern emerging in the dynamics. The process jumps between three phases, two of which are strongly periodic, suggesting two or more attractors.

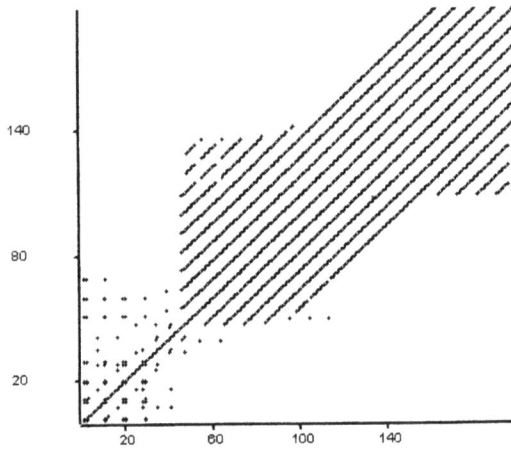

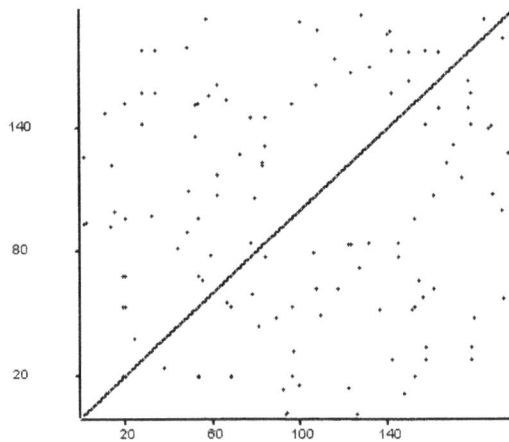

Figures 6.17 and 6.18: Recurrence plots for Tribmod 19 and Tribmod 19θ

The effect of θ randomization is to destroy both the broken and the continuous 19 periodicity. The initial period of instability is much shorter here.

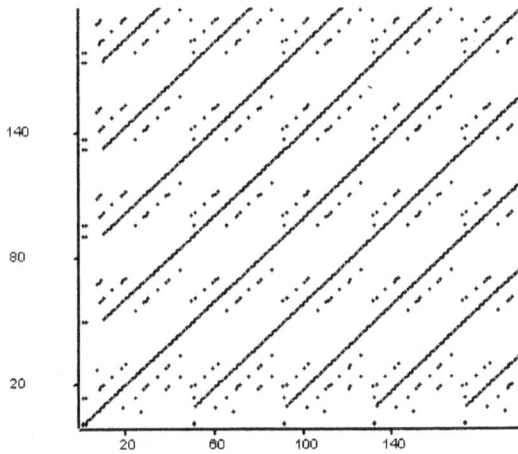

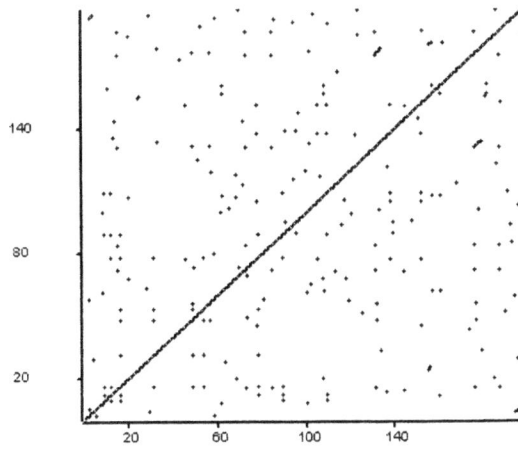

Figure 6.19: Recurrence plot with narrow noise on k, k = 7 and 10

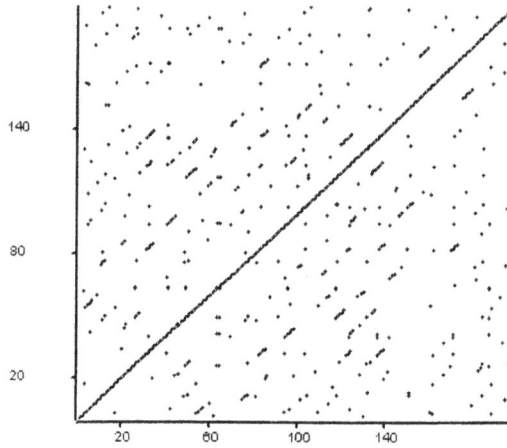

Figure 6.20: Recurrence plot with narrow noise on k, k = 7 and 10

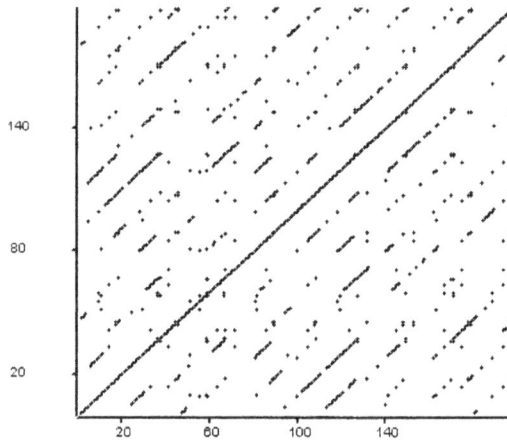

Figure 6.21: Recurrence plot with narrow noise on k, k = 13 and 19

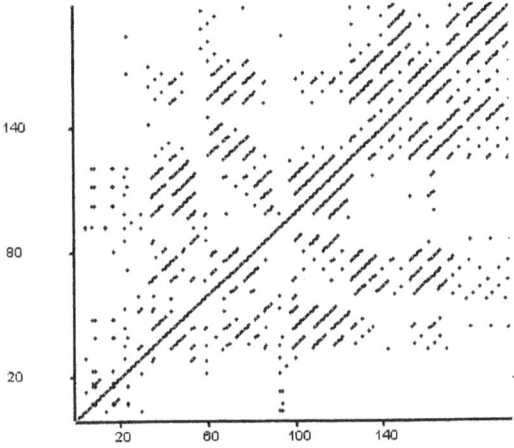

Figure 6.22: Recurrence plot with narrow noise on k, k = 13 and 19

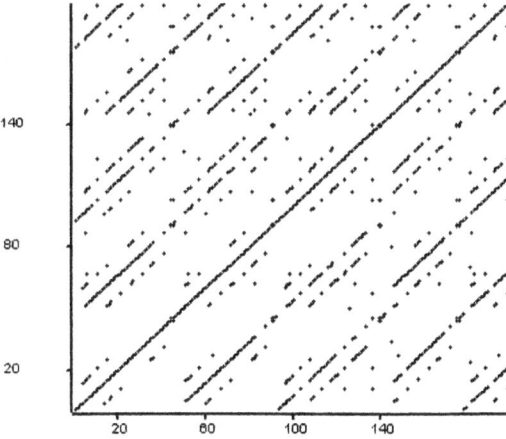

Figure 6.23: 2-d Delay plot for Tribmodk

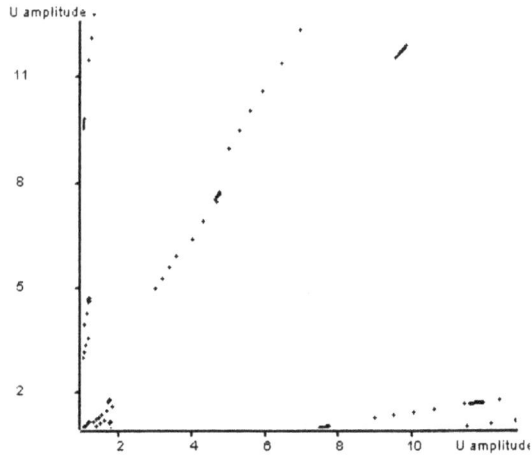

Figure 6.24: 2-d Delay plot for Tribmodkθ, k=13

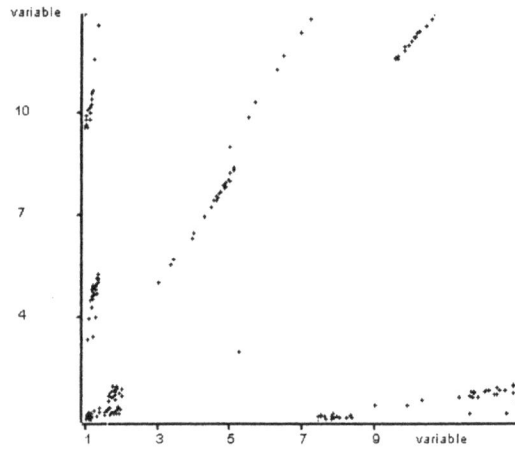

Figure 6.25: 2-d Delay plot for Tribmodkθ, $\theta \sim rect[k \pm 1]$

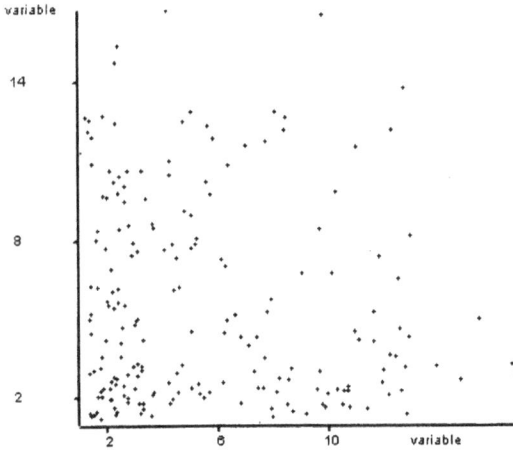

Postscript on the Figures

The marked diversity of the recurrence and delay plots induced by small changes in the parameters is not unique to the Tribonacci series and can be illuminated by recent discussions of the problem of defining exactly what chaotic means. The most-frequently quoted definitions or empirical tests for the presence of chaos include sensitivity to initial conditions and positivity of the largest Lyapunov exponent, though neither are strictly necessary. Sobottka and de Oliveira (2006) draw a distinction between periodicity and eventual periodicity, trajectories that begin aperiodicially and only after evolution become observably periodic are a property of chaotic dynamics, so recurrence plots that appear random in early iterations and then periodic are to be expected.

Another feature that Yao, Yu, Essex and Davison (2006) consider fundamental and a definitional property is mode competition, chaotic attractors create at least two internal and competing modes. These may be revealed in a frequency analysis, and in unusual patterns in delay plots.

Chapter 7

Rescorla's Theory of Conditioning

Rescorla and Wagner (1972) published an influential model of classical conditioning, which has been further refined by Rescorla (2001), introducing as necessary an ogival or cubic function which departs from its original linearity. The question of interest is whether this revised model could be subsumed as a special case of nonlinear psychophysical dynamics. The original model, being linear and deterministic, was widely studied and modified in various ways to try and incorporate stochastic ideas and thus cope with the probabilistic nature of competing response behaviour and response measures (Frey & Sears, 1978; Hanson & Timberlake, 1983).

Alternative models for interaction effects between different cues in making causal judgements have evolved that vary both in their structure and the number of their free parameters, as well as in prior assumptions about what entities may be present in a decision situation. White (2005) gives a review and it is known that effects can be predicted by models better for this purpose than Rescorla's. White's algebra is basically built from linear weighted sums of dimensional components and is extended to cover 24 different conditions for combining evidence. He contrasts it with some Bayesian nets used to model decisions made by children (Gopnik,

Sobel, Schulz, and Glymour, 2001). We will not here go beyond considering the Rescorla algebra. Its use in causal learning is reviewed briefly at
the end of this chapter.

We will keep Rescorla's notation for his model throughout:

For a single stimulus

$$\Delta p_n = \beta(\lambda - p_n) \qquad [7.1]$$

where β is the learning rate parameter, p_n is the probability of a response
on trial n, and λ is the asymptote of learning. The complications begin
when compound stimuli, made up of two or more components, are used.

The model is now expressed in terms of V_i, which is the strength of
association to stimulus i. V may be positive or negative and is thus not a
probability. The compound stimulus is called AX, and V_{AX} is the strength
of association of the compound stimulus. The linear assumption is, then,
that

$$V_{AX} = V_A + V_X \qquad [7.2]$$

which appears from more recent data to be false. When a stimulus compound was followed by a US, the changes in the strength of each of the
components A and X were taken to be a function of V_{AX}. This was noted
to be an important departure from linearity (Rescorla & Wagner, 1972, p.
76), but different from that in [7.2].

When a compound AX is followed by US_1 then

$$\Delta V_A = \alpha_A \beta_1 (\lambda_1 - V_{AX}) \qquad [7.3]$$

and

$$\Delta V_X = \alpha_X \beta_1 (\lambda_1 - V_{AX}). \qquad [7.4]$$

where $0 \leq \alpha, \beta \leq 1$. The α are scalar weights. If a different US_2 is employed, then the suffices 1 are replaced by 2 in all cases in [7.3] and [7.4].
It is assumed that λ is a function of the magnitude of the US. There are
some vaguenesses or, to be more charitable, a lack of formal specification
in the model concerning the boundedness of λ and the Vs. A monotonic
mapping of V onto response probabilities was assumed.

It follows that, when both V_A and V_X begin conditioning at zero, that after any large number of conditioning trials with AX reinforced

$$V_X = \frac{\alpha_X}{\alpha_A + \alpha_X} V_{AX} \qquad [7.5]$$

which resembles Luce's version of Individual Choice Behaviour (1959), now known to be in general false.

Extension to Discrimination Learning

Suppose that in a series of trials where AX and BX are mixed randomly, AX is always followed by reinforcement and BX is always followed by nonreinforcement, to induce a discrimination between A and B. Here X can be thought of as a background or as contextual cues.

$$\begin{aligned} \Delta V_A &= \alpha_A \beta_1 (\lambda_1 - V_{AX}), & [7.6] \\ \Delta V_X &= \alpha_X \beta_1 (\lambda_1 - V_{AX}) & [7.7] \end{aligned}$$

where AX is reinforced, and

$$\begin{aligned} \Delta V_B &= \alpha_B \beta_2 (\lambda_2 - V_{BX}), & [7.8] \\ \Delta V_X &= \alpha_X \beta_2 (\lambda_2 - V_{BX}) & [7.9] \end{aligned}$$

where BX is nonreinforced. Together, [7.6, 7.9] is a nine-parameter model; simplifying assumptions by setting all αs = 1.0 and βs = 0.5, and Vs zero prior to the first learning trial were made. The main point of interest is that AX increases monotonically to an upper asymptote, but BX can variously rise and then fall, or fall from the onset of learning. Heath (1979) found the linear model inadequate and augmented it using stochastic decision processes.

Various probabilistic schedules of reinforcement are used in conditioning, and predictions incorporating π, the reinforcement probability on a trial, are derived. Consequentially, π is also the proportion of reinforced trials in a closed sequence. The asymptotic value of partially reinforced (i.e. $\pi \simeq 0.5$ for each compound stimulus) V_{AX}, V_{BX} is

$$V_{asympt} = \frac{\pi \beta_1 \lambda_1 - (\pi - 1)\beta_2 \lambda_2}{\pi \beta_1 - (\pi - 1)\beta_2} \qquad [7.10]$$

anf if $\lambda_1 = 1.0, \lambda_2 = 0$ and $\pi = 0.5$ then

$$V_{asympt} = \frac{\beta_1}{\beta_1 + \beta_2} \qquad [7.11]$$

Rescorla and Wagner derived a diversity of expressions for asymptotic behaviours under assumptions about parameter values and using the linearity assumptions $V_{AX} = V_A + V_X$ and $V_{BX} = V_B + V_X$. An attempt to circumvent criticisms of the linearity assumption was suggested by treating the stimuli involved as sets V_{AX}, V_{BX}, V_{ABX} because $V_A \cap V_B \neq 0$. Then $V_{ABX} \rightarrow \lambda_1$ and $V_{AX} \rightarrow \lambda_2$ and so for V_{BX}, so that $V_X \rightarrow (2\lambda_2 - \lambda_1)$, with $V_A, V_B \rightarrow (\lambda_1 - \lambda_2)$.

Those details constitute a sufficiency for considering if a parallel with Γ models in NPD is viable.

Structural Analogies

The conditioning theory is essentially about short trajectories leading to asymptotic values where the asymptotes are relative response probabilities. In a sense, it is about the orbits of the attractor manifold of the dynamics, but the model's only dynamical equations are [7.1, 7.3, 7.4]. The data and theory in the source papers do not usually show confidence limits on predictions or data, although these could be presumably be derived with ancillary assumptions about binomial distributions on the p_n.

The entry points for comparison are the equations for two-component mixtures in each theory, in conditioning [7.3, 7.4], and in NPD (Gregson, 1992, p. 24, eqns [2.19, 2.20]) for two continua h, i, (where $i = \sqrt{-1}$ when it is a multiplier and not a suffix)

$$Y_{h(j+1)} = -a_h \cdot (Y_{hj}^* - 1)(Y_{hj}^* - i\lambda_h \cdot (a_i)^{-1})(Y_{hj}^* + i\lambda_h \cdot (a_i)^{-1}) \quad [7.12]$$
$$Y_{i(j+1)} = -a_i \cdot (Y_{ij}^* - 1)(Y_{ij}^* - i\lambda_i \cdot (a_h)^{-1})(Y_{ij}^* + i\lambda_i \cdot (a_h)^{-1}) \quad [7.13]$$

Both models are written in difference equation form and both can induce local transient nonmonotonic trajectories in their dependent variables, the comparisons would have to be with Vs in [7.1, 7.3, 7.4] and Y(Re,Im) in [7.12, 7.13]. What is expressed as two stimuli, A, B, and a background, X, in pseudodiscrimination experiments is expressed as two complex variables, Y(Re,Im), with shared Y(Im) in 2D NPD **Case 2** theory. The fundamental difference in the objectives of the two theories is that one was written for learning and the other for sensation. The possible comparison arises when responses to mixtures are mediated by nonlinear interactions between continua.

Parameter Comparisons		
Conditioning	Psychophysics	Meaning
p_n	[Y(Re)]	response probability
β	[a]	learning rate
α	[e]	weight on β
π	[Y(Im)]	reinforcement probability
λ_1, λ_2		asymptotes on V
V_A	[Y(Re)]	associative strength of A
V_{AX}	[Y(Re,Im)]	strength of A with noise X
V_{AB}		strength of mixed output AB
[β_1, β_2]	a_1, a_2	gains on each dimension
[a]	e	sensitivity
[V_A]	$0 < Y$(Re)< 1	observable output (response)
[π]	Y(Im)	internal activity
[V_{ABX}]	λ	cross-coupling
	η	delay time

Note in the table that λ means an asymptote in conditioning theory, but means a cross-coupling in NPD. The two meanings are quite different, but the symbolism has been retained, so that reference to the original sources can be made simpler. We are going to drop the asymptote idea later in revising NPD to incorporate some sort of conditioning process.

To compare the two models, the parameters have to be matched, if they functionally correspond. Such correspondence does not necessarily have to exist for all parameters; that is, their role within the model may be given the same meaning, but obviously their numerical values do not match. As NPD is written as a difference equation, [7.1] may be revised to facilitate the explication and comparison of model differences as

$$V_{n+1} = (1 - \beta^*)\lambda + \beta^* V_n \qquad [7.14]$$

where $\beta^* = 1 - \beta$. Equation [7.14] shows that the model is a linear weighted compromise of a postulated asymptote and the present value of V. There is no need to put an asymptote into Γ [7.12,7.13] because that recursion is self-limiting in $Y(\text{Re})$ within its attractor basin of stability. In the table of Parameter Comparisons near-functional equivalences have been indicated in [..]; such equivalences are not necessarily symmetrical.

The Γ recursions are the basis of the observable map $M_\Gamma := a \mapsto Y(\text{Re})$ and the conditioning of the map $M_{cond} := \pi \mapsto V$, as a and π are under experimenter control and functions of $Y(\text{Re})$ and V are treated as response scale values. As M_Γ is the result of a complex cubic polynomial function, it is ogival. Rescorla (2001) has now revised M_{cond} to be ogival because a linear map does not fit data. One may say that end effects at zero or near asymptote exhibit nonlinearities; this is a weak convergence of the two models. The question we now explore is: can the conditioning model be rewritten as a special case of NPD? It could be argued, on the basis of a sort of biological economy, that the same brain has to support the sequential dynamics of the domains of activity, sensation and learning. Any model which is linear in its details can be treated as a local subregion of the dynamics of a more general nonlinear model on the principle of piecewise linearization.

The entry point of the argument is to note that $\Delta Y = f(e, Y(\text{Im}))$ and $\Delta V = g(\alpha, \beta)$ are bivariate operator equations f, g respectively in the two models that control rates of change of their dependent variables. But in conditioning there is an extra variable π, a variable reinforcement parameter, that has no counterpart in sensation. This is equivalent dynamically to having a nonstationary feedback loop added to the recursion. To recall the distinctions of the introduction, we move from U to C time series.

A simple heuristic way to parallel the reinforcement parameter in an augmented Γ is to make e in one-dimensional NPD a function of an external schedule S_j.Then

$$\Gamma := Y_{j+1} = -a \cdot (Y_j - 1)(Y_j - i \cdot f(S_j))(Y + i \cdot f(S_j)), \ i = \sqrt{-1} \quad [7.15]$$

The time series generated by [7.15] is

$$Y_{j+1} = f(a, S_j) \qquad [7.16]$$

and [7.14] with π incorporated as in [10] is

$$V_{n+1} = (1 - \beta')\lambda + \beta' V_n \qquad [7.17]$$

where now $\beta' = (1 - \pi\beta)$.

The suggested necessary modification of [7.2] which Rescorla (2001, p. 64) considers for compound stimuli is

$$V_A AB = f(V_A) + f(V_B) \qquad [7.18]$$

and now this f is sigmoidal, so it would resemble $M_\Gamma{}^1$.

---

[1] The theory variant called $\Gamma V1$ in Gregson (1988) uses $e_j = f(\Delta^1 a_j)$, which means that response sensitivity is a function of a local rate of change of stimulation. This is a sort of feedback, whereas π in [7.10] is not contingent on ΔV. To modify [7.15] to accommodate schedules of reinforcement S_j is therefore not the same as using $\Gamma V1$.

Simulation in 2D NPD

The problem now is to see if variations in the parameter values of 2Γ **Case 2** equations can produce some of the same time series for reinforced and unreinforced components of a stimulus mixture. This will be attempted in two stages: first, a purely deterministic treatment of $S = f(j)$ in [7.16], and then a stochastic revision in which the shifts in parameters from trial to trial are probabilistic and not a monotone function of time.

Deterministic model

The qualitative pattern it is desired to reproduce, without any structural changes in the 2Γ model, is that in which the reinforced component slowly rises in output strength with successive reinforcement, but the unreinforced component first rises a little and then falls off to a low asymptote. This rise and then fall in a component which is to be extinguished is supposedly paradoxical for some linear models. It is necessary first to show that a configuration of 2Γ parameter settings which are not implausible, given the extensive results (Gregson, 1992, 2001) in other sensory (but not learning) situations can yield the apparent paradox.

The parameters in a 2Γ **Case 2** model are set initially as $Y_0(Re, Im)_1, Y_0(Re, Im)_2, a_1, a_2, \lambda_1, \lambda_2, \eta$. Putting $Y_0 = (0.5, \eta)$ for both dimensions sets the process as initially stable. $\eta = 10$ has been fixed throughout, as response delay is not a variable system parameter being modelled. $a_1 = 2.4, a_2 = 3.6$ assumes that when the response to dimension 1 is initially unconditioned then the noise strength a_2 is greater. Both the a values are left unchanged, as the actual stimuli do not change, only the relative responses to them. The only parameters available to represent an altered sensitivity to stimuli and interaction between stimuli are λ_1 and λ_2. Continuing reinforcement over a long series $J = 1, ..., N$ of trials thus implies here that $S_J = f(\lambda_1, \lambda_2, J)$.

After some exploratory analyses, it is found that setting initial values $\lambda_{0,1} = 0.2$, and $\lambda_{0,2} = 1.15$, and incrementing on each trial

$$\Delta^1\lambda_1 = +.04, \ \Delta^1\lambda_2 = -.04 \qquad\qquad [7.19, 7.20]$$

yields Figures 1 and 2. Bounds on e_1, e_2 of $e_1 < .45, e_2 > .01$ were set[2], though these are not hit for e_2 in the series until $J = 30$.

Figure 7.1: Time course over trials J of Y(Re)1; reinforced

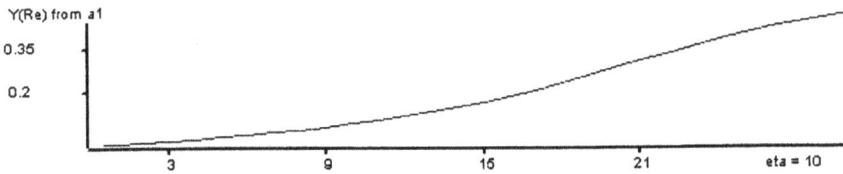

Figure 7.2: Time course over trials J of Y(Re)2; extinguished

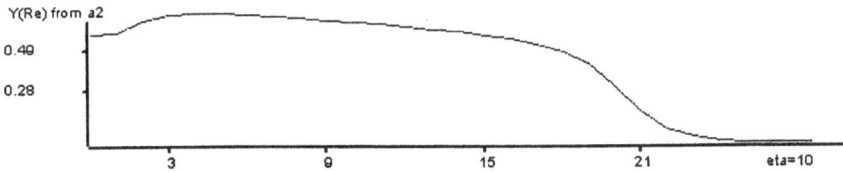

Figure 1 is an analogy of V_A time series and Figure 2 is an analogy of V_X. If an analogy of [7.2] is required then the two curves are summed at any point J. This would here yield an almost flat curve for the mixture until a_1 dominated, which is implausible unless weighted summation were to be used.

---

[2] In **Case 2** the cross-coupling is effected by making $e_1 = \lambda_1/a_2$, and $e_2 = \lambda_2/a_1$. It is known from previous psychophysical results that plausible values of e lie in the range $.01 < e < .45$.

Stochastic models

Now the increments [7.16, 7.17] in $\lambda_{1,\ or\ 2}$ are made with associated probabilities, which adds another two parameters $\pi_{1,\ or\ 2}$ to the model, compare [7.10]. There are obviously various ways in which this can be done; one, which creates slight second-order changes to the Figures 7.1 and 7.2, is to write the random variable $\kappa \sim RECT(0,1)$, and then

$$\text{for reinforcement: } \lambda_J = \lambda_{J-1} + \kappa \cdot .2 - .05 \qquad [7.19]$$

$$\text{for extinction: } \quad \lambda_J = \lambda_{J-1} - \kappa \cdot .2 + .05 \qquad [7.20]$$

which produces Figures 7.3 (corresponding to 7.1) and 7.4 (corresponding to 7.2). The shifts to conditioning or extinction now arise sooner and more abruptly and the reinforcement curve runs to an asymptote.

Figure 7.3: Time course of reinforcement with stochastic perturbation

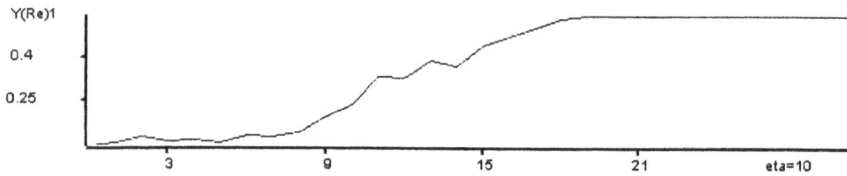

Figure 7.4: Time course of extinction with stochastic pertubation

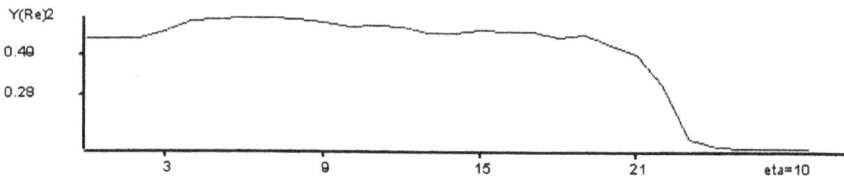

Note that the initial rise in the extinction curve still exists. This rise exists if there is an appropriate choice of a combination of all of the parameters a_1, a_2, λ_1 and λ_2.

Another much more erratic substitution of a stochastic component on the λ values is shown, without any commitment to its plausibility. When conditioning experiments are run, there is no certain smooth improvement in performance, and local failures or relapses can arise, but unless these are reported in detail without averaging over replications (which is not usual) it it not possible to know if they are faithfully modelled in any simulation. However, the next two figures 7.5 and 7.6 depict what can happen. The equations corresponding to but replacing [7.19,7.20] are ($J = 1, ..., 30$)

$$\text{for reinforcement: } \lambda_J = \lambda_{J-1} + .04 \cdot J(2.0 \cdot \kappa - 1.0) \qquad [7.21]$$

$$\text{for extinction: } \quad \lambda_J = \lambda_{J-1} - .04 \cdot J(2.0 \cdot \kappa - 1.0) \qquad [7.22]$$

Figure 7.5: Reinforcement with erratic stochastic effects

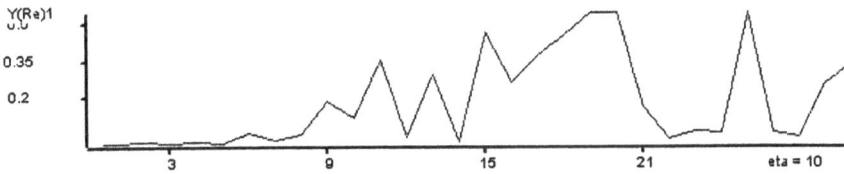

Figure 7.6: Extinction with erratic stochastic effects

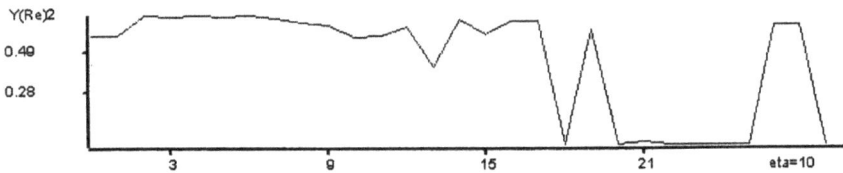

Higher and Indeterminate Dimensionality

The Rescorla-Wagner model has been used in other contexts than the original conditioning paradigms for which it was created. One area of interest is the learning and use of causal inference in humans as a topic in developmental psychology (Gopnik, Glymour, Sobel, Schulz, Kushnir and Danks, 2004). It transpires that the Rescorla-Wagner model can not capture some processes of causal inference, but another approach, based on directed acyclic graph theory, can do better. Such graphs are called Bayes nets when combined with a recursive strategy to choose between various links between nodes. With the subsequent use of Bayes theorem to compute various posterior probabilities associated with different links and subsets of nodes in the graph. Bayes nets were developed in artificial intelligence as one way of solving what has come to be called the inverse problem (Pearl, 1988). This finding about the inadequacy of the Rescorla-Wagner model is pertinent here because, to the extent that the Rescorla-Wagner equations can be treated as analogous to NPD equations, then those also would fail if used without modification as suitable for representing inference judgements.

There are, however, some complications: in equations [7.1-7.22] we have restricted comparisons to what are two-dimensional processes and causality situations generally involve more terms, so that directed paths like $X \rightarrow Y$ may be wrong as failing to reveal that the true situation is $X \perp Y, Z \rightarrow Y, Z \rightarrow Y$ where $\perp$ means causally independent. This is like the situation in statistics where partial correlations and not first-order correlations should be used to reveal psychophysical influences like those mapped in Chapter 1, in Figure 1.3.

Gopnik et al, (2004,p.6) observe pertinently that

> Causation is not just correlation, or contiguity in space, or priority in time or all three. Causal structures rarely just involve one event causing another. Instead, events involve many different causes interacting in complex ways. A system for recovering causal structure has to untangle the relations among those

causes and discount some possible causes in favour of others.

Even if the underlying causal relationship is deterministic, the occurrence of other causal factors, which may not be observed, will typically make the evidence for the relationship proba- bilistic.

All this is true; the a priori indeterminacy of the number of variables, and hence the dimensionality of the system, is critical and almost always greater than two, but it is still not sufficient. The Bayes net approach is acyclic, so there are no feedback loops in it, and the assumptions are that deterministic and stochastic linkages are all that may prevail; there is no provision for some of the characteristic features of nonlinear dynamics, including mixes of fast and slow dynamics, which arise naturally in NPD but not in R-W. White (2005, p. 131) suggests that the RW model can be supplemented by "an account of within-event associations", which would involve more parameters and something like adding fast dynamics. At the same time, it is clear that causal learning involves more than bare psy- chophysical mappings, that are necessary but not sufficient.

Both Rescorla-Wagner equations and NPD create trajectories of associ- ation or estimates of causal strength; they involve changes over time and run asymptotically to equilibria, whereas the Bayes nets need augmenta- tion to do that. Gopnik et al (2004, p.19) show experimentally that there are problems children can solve, involving what is called "backward block- ing", that are outside the scope of causal Rescorla-Wagner models but can be handled by some Bayes nets. Backward blocking is the situation where learners decide if some event is a cause of an effect by using information where the event never appears in conjunction with other trials where it does appear. The RW model can be modified by adding axioms that de- crease the association of a cue with an outcome when the outcome occurs in the absence of the cue. Also, the RW model requires a prior specification of the direction of putative causal links, whereas Bayes nets (and humans) can, in some circumstances, decide between $X \rightarrow Y$ and $Y \rightarrow X$. It may be important that in using $n\Gamma$ equations we have always that

$$Cause \rightarrow a, \ Y \rightarrow Effect,$$

that is, causal direction is expressed in model structure in a different way from that in RW. If the extension to $n \times n\Gamma$ is used (Gregson, 1995), then the structure of an $n \times n$ matrix of λ coefficients in [7.12, 7.13] is also needed to be solved for. That corresponds to restricting the (0,1) links in a directed Bayes net.

At the present state of knowledge, it can be safely asserted that none of the competing models have universal validity for explaining how and what causality inferences are actually made by human subjects. Conditioning theory, of which RW is a particular case, does not effectively deal with human contingency judgements (Shanks, 1985), nor with the role of associated implicit stimuli (Miller, Barnet and Grahame, 1995).

Chapter 8

Nonlinearity, Nonstationarity and Concatenation

A comparison of six time series two from pseudo random generation, two from convoluted theoretical psychophysics, and two from EEG records are compared using a set of four statistics. It is seen that local largest Lyapunov exponents, the entropic analogue of the Schwarzian derivative, higher-order kernel matrices, surrogate random tests for confidence limits on parameters and eigenvalues of the dynamics all yield different information about the local instabilities of the processes. All the time series are, in some way, different from one other.

> Only something that appears to be both orderly and disorderly, regular and irregular, variant and invariant, constant and changing, stable and unstable, deserves to be called complex.
>
> (Edelman & Tononi (2001, p. 135).

The investigation of unstable quasi-periodic orbits where there is noise present and surrogates may be employed has become of increasing interest in the study of biological data. Such data include psychophysiological time series such as scalp EEGs (Ding, Ditto, Pecora & Spano, 2001). There has also emerged a diversity of methods for looking at nonlinear nonstationary dynamics in time series, which, however, are only tractable for very long series. For example, Casdagli (1997) reviews the use of recurrence plots on series between 5,000 and 60,000 iterations long. The ESf methods used here were deliberately created to look at very short series from psychophysical and psychophysiological data sets, where 100 iterations under stability is often a practical limit. The general problems of employing statistical analyses to discriminate between randomness and chaotic dynamics in time series, particularly in higher dimensionalities, something that is not always possible, are reviewed by Berliner (1992) and commentators and the valuable role of symbolic dynamics was then already known.

In psychophysiology, the identification of transient nonstationary patterns of activity has been explored theoretically as a basis for perceptual and memory processes (Gregson, 1993, 1995; van Leeuwen & Raffone, 2001). The idea being pursued is that the dynamics of neural activity have to be very labile for the brain to be able to learn and to revise rapidly and continually its processing of perceptual inputs. In time series terms, this lability implies possible nonstationarities and singularities. So, except over very brief realisations, statistics such as frequency domain spectra or autocorrelations will mask and not reveal singularities as they involve only a small part of the energy distribution of the process and become masked in any averaging over time. In the particular case of the Fourier transforms, by FFT analyses, if one were to capture the singularities or abrupt jumps, spikes or bursts, as they are variously called in EEGs, then in the limit one would need an infinite series of frequency terms in the Fourier expansion. The alternative of treating spikes or bursts a priori as outliers is not acceptable in nonlinear dynamics, though outliers can exist as a consequence of identifiable exogenous perturbation.

Thom (1975) noted that, in biological data, there may be many local small catastrophes in the evolution of a process but, because they are many and small, their existence would be smoothed out in modelling. However, if one can take short enough series and explore the dynamics within such epochs, as compared with their smoother neighbouring epochs of similar length, then the nature of the nonstationarities in the dynamics may be clarified. The questions of how and why local anomalous dynamics arise are tied to the identification of fast/slow dynamics. Van Leeuwen and Raffone (2001) come to this problem through considering the activity of the hippocampus as mediated by nonlinear coupled maps. Coupling of fast and slow dynamics can be modelled in various ways: the NPD approach (Gregson, 1998, 1992, 1995) does it with complex variables, and it can also be done with real signals riding on slow carrier waves. In psychophysiology, it is noted that the time scale of sensory processing is much faster that that of memory consolidation, though both may run in parallel and may involve neural circuitry via the hippocampus.

Shafer (1994, p.60), discussing the logic of statistical inference writes

> We deliberately construct the stochastic story that serves as our standard for comparison. We often construct several.....we may (or may not) find a stochastic story in which the performance of the spectator roughly matches the performance of our forecaster At no point are we required to think of the forecaster herself or of the phenomenon being forecasted as part of a stochastic story.

The fundamental difference here is that trajectories which are not stochastic, or are mixed deterministic and second-order stochastic, may be used instead of stochastic stories; the matching idea is still the same.

For illustration, two EEG samples are compared with four processes, two of a complicated nonlinear form, and two which are Gaussian i.i.d. series. Thus there are six series, two from real data and four from theoretical computations. For each series of 960 steps, there are eight subseries of 120 steps. There are four statistical methods used on each subseries, so the total study is $6 \times 8 \times 4$ in its full crossed design. This does not imply that it is amenable meaningfully to Anova.

Statistical Methods

This section recapitulates some previously described uses of methods, to make the immediate contrast of the methods and their derived indices more compact, without back reference to the previous chapters.

Lyapunov Exponents

Under some conditions where the diffusion rate of the dynamics of the time series is nearly zero, the probability distribution of the local largest Lyapunov exponent is approximately log-normal (Fujisaka, 1983) and, under stricter conditions, the difference between the entropy and the largest Lyapunov exponent is approximately constant (Kohmoto, 1988, p. 1349, eqn. 5.9). It is not yet known how well real psychometrical data might satisfy these analytical conditions. See Appendix 1 to this chapter for a definition in terms of Jacobians.

Random Surrogate series

The method of random surrogates to compare a data series with a number of randomised permutations of the series was introduced by Theiler *et al* (1992). It is used here in a slightly different manner, though the logic is comparable. The objective is to match a series in its first two moments, but to remove any sequential dependencies in the surrogates.

Entropic Analogue of the Schwarzian Derivative

This method was introduced by Gregson (2001) and is a parallel of a derivative introduced by Schwarz (1868), but is based on local summations of coarsely scaled series and not on point derivatives of a continuous function. The idea of treating a dynamical trajectory from an entropy perspective is not novel but is well developed (See Sinai, 2000, Chapter 3) and can be traced back to statistical mechanics in the 19th century.

A series of a real variable y is partitioned into k exhaustive and mutually exclusive subranges in the values y it takes, $k = 10$ is initially sufficient. As a condition of the normalisation $0 \leq y \leq 1$ the end subranges $h = 1, k$ will initially not be empty. Some or all of the remaining $k - 2$ subranges may be empty when the dynamics are minimally informative. Call the width of one such subrange $\delta^{(0)}$ (of y). $\delta^{(0)} = .1$ for the original y but will almost always be less for $\Delta^m(y)$, the successive m^{th} absolute differences of the series. This δ is the *partitioning constant* of the system. Call the partitioning constant of the range of the mth differences $\delta^{(m)}$. In the examples used here, $\delta^{(1)}$ is referred to simply as δ in the tables of results, and all $\delta^{(m)}, m > 0$ are set constant $= \delta^{(1)}$.

The frequencies of observations lying in one segment $\delta_h, h = 1, ..., k$ is then n_h, and converting to probabilities p_h we compute the information in that subrange. The absolute differences of the rescaled y series are taken putting

$$\Delta^1(y_j) = |y_j - y_{j-1}| \qquad [8.1]$$

and further differencing repeats this operation, so that

$$\Delta^2(y_j) = |\Delta^1(y_j) - \Delta^1(y_{j-1})| \qquad [8.2]$$

. This operation can be continued only until all absolute differences are zero. Going only as far as Δ^4 is sufficient. Summing over all subranges gives the total information in the mth differenced distribution as

$$I_{\{h\}}^{(m)} = I^{(m)} = \sum_{h=1}^{k} p_h log_2(p_h)|_m \qquad [8.3]$$

Then by definition the entropic analogue of the Schwarzian derivative, abbreviated to ESf, has the form

$$ESf := \frac{I^{(3)}}{I^{(1)}} - \frac{3}{2}\left(\frac{I^{(2)}}{I^{(1)}}\right)^2 \qquad [8.4]$$

For some strongly chaotic (theoretical variants on Γ) series, ESf is positive in the examples we have seen and becomes increasingly negative as processes are more stochastic. It has been applied to series from psychophysics, EEGs, climate and economics. Its main advantage is that is is computable over much shorter time series than the Lyapunov exponents.

Bispectral Kernel Analysis

This method is usually employed in the frequency domain but here a time domain version is used together with the surrogate tests.

Bispectral analyses are, in fact, third-order kernels of time series. They have been extensively used in anaesthesiology, in the tracking of the evolution of EEGs during surgery, following Rampil (1998) and Proakis *et al* (1992), and are there computed in the frequency domain using FFTs. They resemble the kernels used in nonlinear analyses described by Marmamelis and Marmarelis (1978).

If there exists a real discrete zero mean third-order stationary process x, then its third-order moment matrix is defined over a range of lags $1, ..., m$, $1, ..., n$ by

$$R(m, n) := E[x(j) \cdot x(j + m) \cdot x(j + n)] \qquad [8.5]$$

This matrix is skew symmetric, as m, n can be exchanged in the definition. The triangular supradiagonal matrix is sufficient for exploratory purposes and is used in the tables in the form shown. To compute its eigenvalues, it is reflected into the square form and the leading diagonal cells filled with the average of the off-diagonal cells as an approximation. Its roots will, in general, be a mixture of reals and complex conjugates.

$$
\begin{array}{lllll}
b(1,2) & b(1,3) & b(1,4) & b(1,5) & b(1,6) \\
b(2,3) & b(2,4) & b(2,5) & b(2,6) \\
b(3,4) & b(3,5) & b(3,6) \\
b(4,5) & b(4,6) \\
b(5,6)
\end{array}
$$

Compare the usual second-order autocorrelation which is defined as

$$R(m) := E[x(j) \cdot x(j + m)] \qquad [8.6]$$

over a range of lags $1, ..., m$.

The Time Series Used

(1) A Quasi-Random Gaussian almost-i.i.d. Series

This series was generated, using an algorithm RAN2, by Press et al (1986), with only one linear congruential generator. It is relatively coarse in resolution of the numerical values of normal deviates created by subsequent use of a function GASDEV to transform to Gaussian deviates on the rectilinear deviates RECT(0,1) first generated by RAN2.

Figure 8.1: (1) Gaussian pseudo-random series

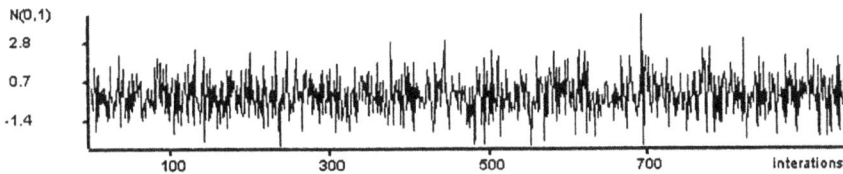

(2) A Second Quasi-random almost-i.i.d. Series

This series was also generated from an algorithm by Press et al (1986), this was RAN1 employing three linear congruential generators. It is supposed to have finer resolution of the values generated and to show no sensible sequential correlations. It is assumed that the correlations in question would be first-order of varying lag, as used to create the autocorrelation spectrum of the series. Its periodicity is far beyond the series length used here. The function GASDEV was again used to create a series of random Gaussian deviates.

(3) Convoluted Gamma Concatenated

Convoluted Gamma trajectories have been previously used for illustration (Gregson, 2000). As they are not phase-locked their short realisations may be concatenated.

Figure 8.2: A second Gaussian pseudo-random i.i.d. series

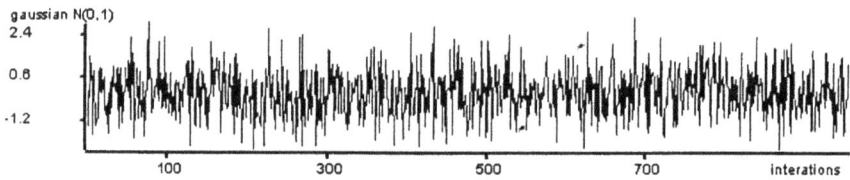

Table 8.1: Statistics of the two Gaussian pseudo-random series
Distribution, trend and autocorrelation parameters

Statistic	series (1)	series (2)
mean	-.035	.030
min	-2.771	-2.390
max	4.379	3.168
variance	.986	.932
skewness	.194	.116
kurtosis	.260	-.117
regr: slope	8.4×10^{-5}	-9.0×10^{-5}
AR(1)	-.0048	+.0191
AR(2)	-.0121	+.0098
AR(3)	-.0231	-.0175
AR(4)	.0168	-.0256
AR(5)	-.0214	-.0604
AR(6)	-.0317	-.0062
AR(7)	+.0148	-.0018
AR(8)	+.0125	-.0171
AR(9)	+.0109	-.0165
AR(10)	+.0250	-.0186

Figure 8.3: The Raw ConvΓConc series

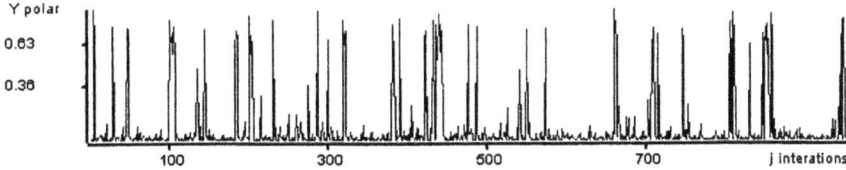

The first stage is to convolute, by having the parameter a in $\Gamma V7$ a variable, with a random bounded rectangular distribution. That is, for a complex variable $Y(Re, Im)$

$$\Gamma := \quad Y_{J,j+1} = -a_J(Y_j - 1)(Y_j - ie)(Y_j + ie), \qquad i = \sqrt{-1},\, j = 1, ..., \eta \tag{8.7}$$

where $J\ (= 1, ..., N)$ is one cycle in the convolution, and

$$a_J \sim \alpha \cdot RECT(0, 1) + \beta, \qquad a_j \sim i.i.d. \tag{8.8}$$

where α and β are real scalars. They have to be small and bounded to avoid the process exploding. An additional complication introduced in this example is that $e = c \cdot \Delta^1(a_J)$; again, c is bounded. The trajectory returned for examination is the polar modulus r value, where

$$Y(Re, Im) \mapsto Y(r, \theta) \tag{8.9}$$

of $Y_{J,\eta}$ over the range $1, ..., N$. Here $N = 120$ and then eight such series have been concatenated. That is to provide a comparison of the patterns found in some EEG samples (Gregson and Leahan, 2001). This ConvΓConc series with $\eta = 10$ has a curious shape with local erratic high spikes and very high frequency basal activity. Unless it is known a priori that the series is a convolution of two processes, each of [8.7] and [8.8] is in itself stationary in its parameters, then externally observed it is sample of a nonlinear nonstationary process trajectory. Its largest Lyapunov Exponent L is positive: LLE = +.0682

The trajectories created from this convolution may be treated as being on the invariant manifold.

Figure 8.4: The Trimmed Cascaded Γ Concatenated Series

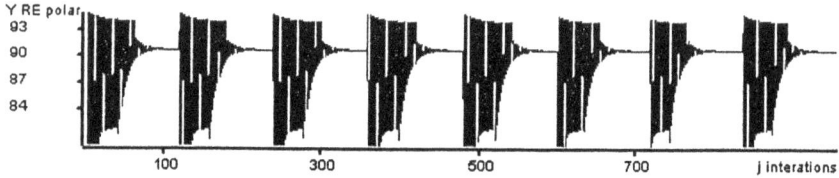

(4) Cascaded, Concatenated and Trimmed Series

Another way of creating a nonlinear and nonstationary series from a Γ seed is to use the linkage

$$a_{J+1} = f(Y_{J,\eta}) \qquad [8.10]$$

and to introduce a second-order random perturbation to each a_{KJ}. Here $J = 1, ..., 150$, $K = 1, ..., 8$. This sort of series is wildly oscillatory at the start of each KJ cycle, and runs into low amplitude period 2^d oscillations, and eventually onto a point attractor for some parameter settings. It is thus a trajectory on the attractor manifold, not the invariant manifold. Its appearance is extremely irregular, but if it is trimmed to dampen the wildest oscillations can be graphed as shown in Figure 8.4. It is questionable if it has any real biological analogues, though conjecturally it might be used as a model of local acute drug action.

The trimming is made by three boundary conditions applied in sequence,

$$
\begin{aligned}
if\ Y < 0, &\quad then\ Y = -Y &\quad [8.11] \\
if\ Y < c_l &\quad Y = c_l + logY &\quad [8.12] \\
if\ Y > c_u &\quad Y = c_u + logY &\quad [8.13]
\end{aligned}
$$

It is only the shape of this series which is of interest, so the raw series may be rescaled linearly and the constants c_l, c_u adjusted afterwards. The constants c_l, c_u are set to bracket closely the very small oscillations after the process converges towards its limit cycle. It has overall LLE = +.0648, but the local LLE is not readily computable unless a high-pass filter is used as well, after which LLE = +.0675

This trajectory lies on the attractor manifold that is deduced from the process running onto a limit cycle in the intervals between random perturbations. Jumping around like that is not necessarily on the attractor manifold, as the Shil'nikov process also has jumps out of and back into a spiral trajectory in the phase space, but lies on the invariant manifold.

(5) EEG under a Control Condition

These data were provided by Kerry Leahan at ANU, drawn from a long series of recordings of EEGs under a diversity of stimulation conditions. This series was taken from a fifth session, after stabilisation, and there was no extraneous stimulation used.

(6) EEG under a Relaxation Condition

The series (1) and (2), and (3) and (4), all being theoretical, may be thought of as null hypotheses for the two real EEG series (5) and (6). The Gaussian series are what Shafer called a stochastic story. Alternatively, the Gaussian series (1) and (2) may be ignored and the contrast between (3) and (4) which are nonlinear comparison stories in the same manner and (5) and (6) explored to see if the Γ series are in any sense models of the EEG processes, though it is not at all probable that they are.

These EEG relax (6) data were recorded from the same study as the previous series (5), but here there was music played as a relaxing background. In the sense of this study we have aperiodic stimulation of a wide spectral form as a weak forcing function, which is not expected to synchronise readily with any of the dominant frequencies in the resting EEG. In fact, it is seen to be nearest to a coarser resolution random Gaussian series, with high dimensionality.

Figure 8.5: EEG under a control condition, and its recurrence plot

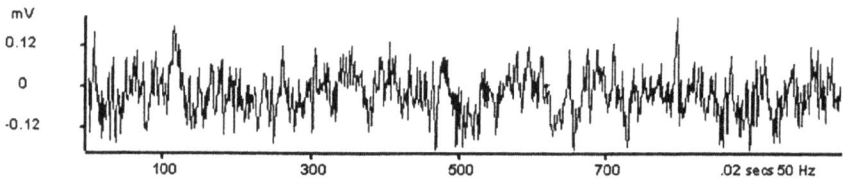

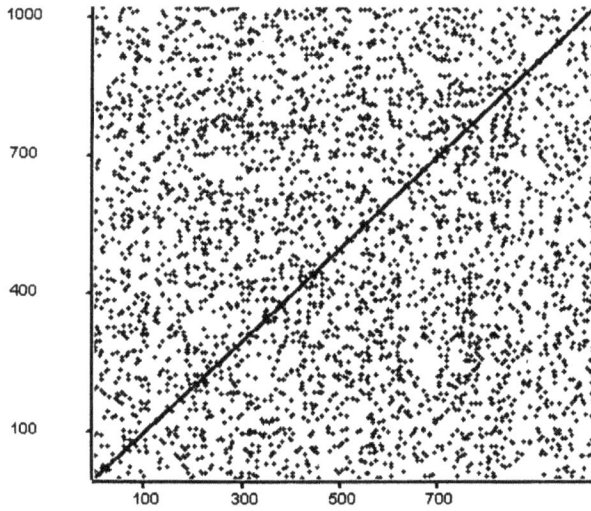

Figure 8.6: EEG under a relax condition, and its recurrence plot

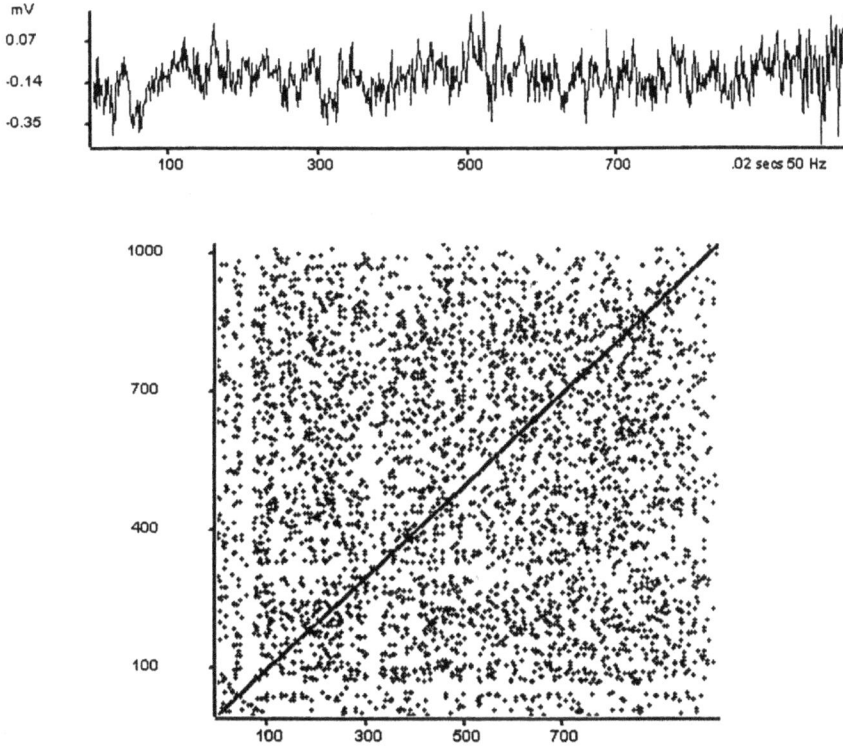

Comparison of Higher-order Analyses

Comparative analyses are given for the six series as a whole and for ESf as partitioned into subseries, at the end of the chapter in Appendix 2.

The tabulations of the ESf and triangular matrices of the kernal ESf values for the two concatenated Gamma examples (3) and (4) reveal consistent differences, even though both are interpreted as chaotic if the LLE positivity is taken as a sufficient statistic.

Table 8.2: Summary of LLE and ESf for whole series

series	LLE	min ESf	mean ESf	max ESf
(1)Gaussian i.i.d	+.1386	-.8435	-.6261	-.3680
(2)Second Gaussian i.i.d.	+.1382	-.6608	-.5809	-.4892
(3) CΓC	+.0682	-.2822	-.1607	-.0187
(4)CCΓT	+.0675	+.3987	+.4330	+.4534
(5) EEG control	+.1268	-.5877	-.4983	-.3501
(6) EEG relax	+.1086	-.8353	-.6316	-.4543

Even though the LLE values are both positive and not very different numerically, the ESf values for both the series and the higher-order kernel terms are very different. For the first case, ConvΓConc (3), the ESf values are small and negative for each of the subseries and in subseries 3 and 6 lie within the surrogate 95% confidence intervals. These are the two subseries which have **n: 6** and have turning points in the local LLE plots. The actual numerical values in the triangular matrices are mostly positive and show some correlation between corresponding cells in the various subseries matrices, which suggests that the process is indeed nearer to chaotic than to random, but shows the nonstationarity which has been observed in the EEG data samples (Gregson and Leahan, 2003). The kernels are, however, very different from those of the EEG series.

The second case, of the trimmed series (4), has a reverse pattern, in that now the ESf values for the subseries are positive and much larger and the two ranges of values do not overlap. None of these ESf values are within their random surrogate confidence intervals. The kernel values are mostly negative, but the **n:** values are consistently lower. The total series is thus a set of trajectories in the attractor manifold, varying only slightly from one subseries to the next. That is, there is a periodic but not sinusoidal forcing function.

So, here are two cases of nonlinear nonstationary series, but their internal dynamics are very different.

Another summary description is provided by the relative frequency of A, W, and B codings for the ESf values for raw subseries in the six tables 8.6 to 8.11 in Appendix 2:

Table 8.3: Summary of A,W,B codings for the ESf values

series	A	W	B
(1) Gaussian	4	0	4
(2) Second Gaussian	5	2	1
(3) CΓC	5	2	1
(4) CCΓT	8	0	0
(5) EEG control	7	1	0
(6) EEG relax	1	4	3

Table 8.4 Eigenvalues for submatrices #1 through #4 in the EEG Control condition (5)

Eigen	Submat # 1	Submat # 2	Submat # 3	Submat # 4
[1]	-3.5813	-2.7014	-3.9175	-3.7362
[2]	+.4180	+.2462	-.5021	+.4131
[3]	-.3201	-.1562	+.4633	-.4002
[4]	-.1103	-.1120	+.3387	+.2502
[5]	+.0781	+.0855	-.2041	-.2299
[6]	-.0421	-.0524	-.0681	-.0164

**Table 8.5 Eigenvalues for submatrices #5 through #8
in the EEG Control Condition (5)**

Eigen	Submat # 5	Submat # 6	Submat # 7	Submat # 8
[1]	-3.6163	-4.0878	-2.5363	-3.7416
[2]	+.3699	+.3576	+.2645	-.3302
[3]	-.2793	-.2719	-.2528	+.2811
[4]	+.2077	-.1576	-.1751	-.2409
[5]	-.1487	+.1305	+.1194	+.2068
[6]	-.1243	-.0424	+.0519	+.0995

In Tables 8.4 and 8.5, the root locus coordinates of the polynomials of the symmetric 6×6 $b(m, n)$ ESf submatrices of the EEG Control data are shown. By the usual mapping, roots in the left-hand half-plane are stable and in the right-hand half-plane are unstable. Complex conjugate pairs, if they exist, denote oscillating components. When both stable and unstable roots are simultaneously present, this is sometimes taken as a necessary but not sufficient sign that chaotic dynamics may be involved.

The dynamics of these matrices differ, particularly in the size of the largest root and the sign of the second largest root. The signs of the fourth and fifth eigenvalues suggest reversals around # 3 and # 6, which could imply local instability in the region of the turning point in the local LLE. Gregson (1984, p.108) gives graphical examples.

It can be seen that the irregularities in the local LLE, the surrogate deviations within the triangular matrices and the root locus patterns are all related ways of identifying transient nonstationarities in the evolution of the dynamics. These local phenomena would be completely masked if an FFT analysis of the whole series were to be used.

Discussion

A distinction should be drawn between nonstationarity and stability. Stability is an assumption in mathematical modelling in various disciplines (Thom, 1975) and involves both the process being modelled and the model itself. It is thus an issue as much in philosophy of sciences as in applied mathematics. Stationarity is a local problem of determining how complex a model necessarily needs to be to represent the evolution of a dynamics process. Thus, a process which is nonstationarity in terms of one model may be stationary in terms of another and then treated as stable over situations to which it may legitimately be generalised. Whether dissipative biological processes which eventually must die can be regarded as stable over more than short epochs is an open question, though Thom took the stance that stability applied to both physical and biological models.

If we compare the various measures, particularly focussing on the short epochs where the **n:** values are higher, there is some weak mutually compatible relationship between the local LLE, the **n:** and the patterns within the recurrence plots for the two EEG series. The actual location in a long series of such epochs may be apparently randomly distributed, but the nonlinear dynamics within these short epochs are compatible with a jump in the phase space of the dynamics.

The results of this and the previous associated study (Gregson & Leahan, 2003) enable us to reiterate some cautions:

(i) The use of single indices such as the LLE is not diagnostically sufficient to identify dynamics which are either not stationary or not stable

(ii) There is a conflict between the need for long time series to fix parameter estimates, which require the calculation of delay coefficients, and the need to describe transient dynamics in local epochs, which naturally emerge within many trajectories, both deterministic and contaminated with second-order stochastic perturbation.

(iii) If a local epoch is in the limit associated with one or more singularities, then the use of spectral analysis, in the frequency domain by methods such as FFT or in the time domain by autoregression, is invalidated in that region and hence globally if the epoch is embedded in an unidentified trajectory.

(iv) Biologically-based processes will exhibit irregular dynamics even without the action of exogenous stimulation and no series on its own, without controls to separate out known exogenous sources, will be fully interpretable.

(v) If the basic iteration frequency of a process in discrete time is not known, it is not possible safely to avoid aliasing over some epochs.

(vi) Higher-order statistics can generate apparently paradoxical results if used with surrogate methods but with no other first-order statistics. They are still valuable for investigating local departures from stability, in either the frequency or time domains.

Notes

I am again indebted to Kerry Leahan for use of some of her EEG data, and to Michael Smithson for the use of his *Mathematica* Eigenvalue functions. The remaining computations were done either with SANTIS software or with Fort77 Linux programs written by me.

Appendix 1: Lyapunov Exponents

We quote from Argoul and Arneodo (1986): "To define the Lyapunov numbers, let

$$J_n = [J(x_n)J(x_{n-1}), ..., J(x_1)] \qquad [8.A1]$$

where $J(x)$ is the Jacobian matrix of the map, $J(x) = \partial F/\partial x$, and let

$$j_1(n) \geq j_2(n) \geq .... \geq j_d(n) \qquad [8.A2]$$

be the magnitude of th eigenvalues of J_n. The Lyapunov numbers are defined as:

$$\lambda_i = \frac{\lim}{n \to \infty}[j_i(n)]^{1/n}, \qquad i = 1, ..., d \qquad [8.A3]$$

The Lyapunov exponents are simply the logarithms of the Lyapunov numbers: $\ell_i = log(\lambda_i)$. We have the convention:

$$\ell_i \geq \ell_2 \geq ... \geq \ell_d \qquad [8.A4]$$

when the system exhibits 'sensitive dependence on initial conditions' at least $L = \ell_1$ is positive."

Appendix 2: Higher-order Subseries

The six tables BESf tables following, 8.6 to 8.11, are each made up of indices from eight subseries of 120 steps long. For the first 7 subseries in each, to save space, we show the following:

Little arrows on the left to show the direction of movement of the local LLE, from graphs in the SANTIS package.

A letter, A, W or B, showing if the ESf for that subseries is above, within, or below the range of its random surrogate confidence interval.

The ESf for that subseries, with its 95% confidence intervals, based on 50 random surrogates.

The triangular matrix of the bispectral coefficients, with any that are within their respective confidence intervals shown as italicised. None of those individual c.i. are shown.

The number of cells which are italicised, shown as **n:**

Table 8.6: (1) Gaussian N(0,1) non-i.i.d. series

LLE	ESf	-c.i.	+c.i.	b(1,2)	b(1,3)	b(1,4)	b(1,5)	b(1,6)
1	$\delta =$	.030		-.6026	-.5128	-.3631	-.4502	-.4772
	-.8435	-.679	-.629	-.6349	-.5887	-.5535	-.5560	
	A			-.3475	-.5696	-.6482		
↘				-.5697	-.5027			
				-.5115				n: 2
2				-.5130	-.5919	-.4521	-.5258	-.4522
	-.7450	-.656	-.608	-.6062	-.5009	-.5561	-.5649	
	A			-.4528	-.4906	-.6254		
↘				-..5554	-.5570			
				-.6577				n: 4
3				-.6121	-.5287	-.6392	-.9262	-.7458
	-.5604	-.719	-.663	-.6347	-.6113	-.4014	-.6080	
	B			-.6276	-.9652	-.5009		
↗				-.4018	-.8290			
				-.6273				n: 4
4				-.3578	-.3573	-.4139	-.3681	-.2889
	-.5435	-.657	-.611	-.4338	-.6418	-.3920	-.4608	
	B			-.3578	-.4591	-.5408		
↗				-.5302	-.7140			
				-.6977				n: 5

Table 8.6 continued: (1) Gaussian N(0,1) non-i.i.d. series

LLE	ESf	-c.i.	+c.i.	b(1,2)	b(1,3)	b(1,4)	b(1,5)	b(1,6)
5				-.4925	-.5693	-.4133	-.4649	-.3254
	-.7625	-.746	-.676	-.6309	-.6619	-.6425	-.4575	
	A			-.6094	-.5704	-.6559		
↘				-.5932	-.5924			
				-.6332				n: 2
6				-.6405	-.7108	-.5276	-.3011	-.4424
	-.3680	-.495	-.464	-.4802	-.5430	-.5934	-.4473	
	B			-.6469	-.4605	-.5055		
↘				-.5276	-.4172			
				-.4471				n: 4
7				-.5230	-.5316	-.4002	-.5051	-.4927
	-.5320	-.638	-.592	-.5143	-.5465	-.3943	-.6504	
	B			-.4880	-.7660	-.4685		
↗				-.7025	-.6369			
				-.6658				n: 2

Table 8.7: (2) 2nd Gaussian N(0,1) non-i.i.d. series

LLE	ESf	-c.i.	+c.i.	b(1,2)	b(1,3)	b(1,4)	b(1,5)	b(1,6)
1	$\delta =$	.030		-.5484	-.5818	-.8449	-.3375	-.4769
	-.6608	-.630	-.590	-.3206	-.5050	-.3655	-.6764	
	B			-.6793	-.5234	-.6761		
↗				-.8438	-.8147			
				-.5573				n: 1
2				-.7419	-.5380	-.5342	-1.1846	-.8552
	-.5159	-.648	-.597	-.7613	-.5022	-.4328	-.5408	
	A			-.6612	-.8101	-.4101		
↗				-.8674				n: 2

Table 8.7 (continued): (2) 2nd Gaussian N(0,1) non-i.i.d. series

LLE	ESf	-c.i.	+c.i.	b(1,2)	b(1,3)	b(1,4)	b(1,5	b(1,6)
3				-.7005	-.5655	-.4672	-.6301	-.6285
	-.5484	-.715	-.660	-.5083	-.2516	-.5792	-.5372	
	A			-.6211	-.5697	-.8095		
↗				-.5250	-.6171			
				-.8182				n: 4
4				-.4925	-.5908	-.5354	-.4622	-.4953
	-.5436	-.698	-.638	-.4600	-.5634	-.5827	-.5483	
	A			-.5271	-.4430	-.5159		
↗				-.5595	-.5280			
				-.4441				n: 2
5				-.7185	-.4952	-.6317	-.5259	-.6388
	-.4892	-.619	-.517	-.5274	-.4150	-.4004	-.4181	
	A			-.6037	-.6573	-.4966		
↘				-.5641				n: 2
6				-.5065	-.5261	-.4333	-.4769	-.6535
	-.6065	-.637	-.595	-.3438	-.5685	-.5576	-. 5728	
	W			-.4877	-.4501	-.9503		
↘↗				-.4851				n: 5
7				-.6038	-.5678	-.6016	-.6186	-.5296
	-.6363	-.749	-.683	-.4347	-.4752	-.7061	-.4910	
	A			-.5375	-.6446	-.4018		
↗				-.6544	-.5205			
				-.7269				n: 3

Table 8.8: (3) Convoluted Gamma Concatenated

LLE	ESf	-c.i.	+c.i.	b(1,2)	b(1,3)	b(1,4)	b(1,5)	b(1,6)
1	$\delta =$	.030		.1146	.0798	.1469	.3044	.4208
	-.1750	-.290	-.197	.1290	.1290	.1629	.3343	
	A			-.3647	.1090	.2219		
↘				.1488	.3944			
				.2345				n: 4
2				.2347	-.0032	-.2641	-.2284	-.0012
	-.0833	-.258	-.519	.3343	.1887	.1469	-.2608	
	A			.2815	.1428	.0280		
↘				-.1965	.0827			
				-.2272				n: 2
3				.2286	.4022	.1494	-.2551	.3059
	-.1798	-.223	-.153	.2946	.1877	.2345	-.1958	
	W			.2219	.2345	-.9050		
↗				.1934	.1859			
				.2020				n: 6
4				.1699	.0848	-.6466	.0710	.2656
	-.2313	-.323	-.258	.0725	.1348	.1629	.2345	
	A			.0890	.1743	.2424		
↗				.1179	.1887			
				-.0561				n: 4
5				.1886	-.3163	-.2255	.2567	.2407
	-.2822	-.249	-.186	.4008	.1038	.0589	.0897	
	B			.2946	-.1927	.0725		
↗				-.1510	.0219			
				.2236				n: 4
6				.1877	.1877	.1939	2145	.4202
	-.2213	-.244	-.147	.2126	.2086	.2610	.0758	
	W			.1484	.1502	-.1737		
→				.2610	.1887			
				.1857				n: 6
7				-.0632	.1313	.2946	-.6338	.4008
	-.0187	-.211	-.130	.1551	.0950	.1352	.1799	
	A			-.0457	.4008	.2610		
↘				-.1634	.2946			
				.2946				n: 7

Table 8.9: (4) Cascaded Gamma Concatenated and Trimmed

LLE	ESf	-c.i.	+c.i.	b(1,2)	b(1,3)	b(1,4)	b(1,5)	b(1,6)
1	$\delta =$	.030		-.4027	-.3761	-.3938	-.3044	-.8265
	.3987	-.640	-.559	-.2261	-1.4135	-.1437	-.6006	
	A			-.4572	-.3684	-.1970		
				-.5902	-.9939			
				-.1221				n: 3
2				-.5812	-.1090	-.5671	-.3170	-.2965
	.4534	-.585	-.508	-.2546	-1.8040	-.4000	-.5308	
	A			-.4509	-.2298	-.2517		
				-.3825	-.6831			
				.1450				n: 3
3				-.6211	-.1273	-.5438	-.2149	-.3579
	.4480	-.588	-.510	-.0637	-1.9672	-.3138	-.5803	
	A			-.3579	-.2198	-.2416		
				-.3446	-.5918			
				.1396				n: 0
4				-.6202	-.1391	-.5657	-.3336	-.2965
	.4501	-.583	-.504	-.2907	-1.7512	-.3747	-.5277	
	A			-.4256	-.2349	-.2517		
				-.3210	-.6937			
				.1633				n: 3
5				-.5255	-.4786	-.8767	-.5047	-.6265
	.4498	-.652	-.565	-.2024	-.8833	-.1720	-.5232	
	A			-.2573	-.3323	-.3086		
				-.4566	-.5438			
				-.2477				n: 4
6				-.2758	-.5734	-.7267	-.3603	-.1862
	.4104	-.565	-.476	.1578	-.4721	-.8300	-.0337	
	A			-.3232	-.4955	-.0627		
				-.1545	-.5212			
				-.3582				n: 2
7				-.4193	-.3874	-.1980	-.0614	-.2643
	.4376	-.522	-.455	-.5529	-.4723	-.3920	-.3513	
	A			-.5434	-.1959	-.4299		
				-.3185	-.5752			
				-.8179				n: 3

Table 8.10: (5) EEG control condition

LLE	ESf	-c.i.	+c.i.	b(1,2)	b(1,3)	b(1,4)	b(1,5)	b(1,6)
1	δ =	.030		-.4809	-.5130	-.5031	-.4853	-.7026
	-.4659	-.697	-.643	-.5510	-.8330	-.6315	-.4919	
	A			-.4650	-.4649	-.6892		
↘				-.6193	-.7701			
				-.7097				n: 4
2				-.5165	-.5564	-.4032	-.3904	-.3257
	-.5206	-.754	-.689	-.3961	-.5104	-.5172	-.4319	
	A			-.5520	-.4702	-.4337		
↘				-.5084	-.3999			
				-.3137				n:3
3				-.5707	-.6562	-.6981	-.5213	-.9822
	-.5778	-.764	-.698	-.4779	-1.0653	-.5449	-.5177	
	A			-.6899	-.5441	-.5724		
↘				-.7427	-.6582			
				-.4835				n:3
4				-.5835	-.6462	-.4875	-.6077	-.8413
	-.3501	-.702	-.644	-.6571	-.8966	-.7204	-.6307	
	A			-.4560	-.5690	-.6307		
↘				-.3311	-.6935			
				-.5230				n:4
5				-.5943	-.5220	-.5648	-.8248	-.4566
	-.5877	-.793	-.702	-.4643	-.4269	-.6950	-.6740	
	A			-.4019	-.7060	-.5778		
↘				-.5699	-.7359			
				-.7585				n: 1
6				-.6417	-.6922	-.6352	-.7396	-.6122
	-.4183	-.818	-.708	-.7996	-.6518	-.7162	-.6837	
	A			-.4546	-.7579	-.5415		
↗↘				-.7587	-.5623			
				-.9324				n: 6
7				-.4430	-.3537	-.4849	-.3311	-.4242
	-.4946	-.495	-.462	-.2403	-.4382	-.4840	-.4505	
	W			-.4491	-.4124	-.6066		
↘↗				-.3012	-.4970			
				-.4048				n: 3

Table 8.11: (6) EEG relax condition

LLE	ESf	-c.i.	+c.i.	b(1,2)	b(1,3)	b(1,4)	b(1,5)	(1,6)
1	δ =	.030		-.8179	-.7041	-.6143	-.5440	-.5498
	-.6806	-.844	-.607	-.5612	-.8827	-.5493	-.7232	
	W			-.5275	-.6470	-.7076		
↘				-.4901	-.6115			
				-.7683				n: 2
2				-.3435	-.3428	-.4339	-.5086	-.4778
	-.4543	-1.537	-.330	-.3727	-.5198	-.5154	-.4037	
	W			-.4401	-.3836	-.3835		
↘				4163	-.5784			
				-.3688				n: 7
3				-.5784	-.6952	-.5581	-.5609	-.4763
	-.6561	-.589	-.326	-.7246	-.5549	-.7149	-.8600	
	B			-.5709	-.6755	-.6035		
↗				-.7626	-.6545			
				-.6025				n: 4
4				-.4592	-.6102	-.4473	-.4629	-.5429
	-.7055	-.473	-.198	-.5734	-.5066	-.6442	-.6557	
	B			-.5957	-.4114	-.4587		
↗				-.7779	-.6349			
				-.6855				n: 2
5				-.5009	-.5867	-.5140	-.5715	-.4637
	-.5820	-2.829	-1.240	-.5875	-.5798	-.3824	-.4935	
	A			-.5887	-.7528	-.5498		
↗				-.5025	-.6087			
				-.4585				n: 2
6				-.8670	-.6597	-.7274	-1.1921	-.9999
	-.6152	-.824	-.261	-.7080	-.7187	-.6743	-.6710	
	W			-.6651	-.5874	-.6209		
↘				-.6289	-.7136			
				-.7115				n: 5
7				-.2904	-.5526	-.4101	-.6620	-.4050
	-.8353	-.667	-.205	-.3986	-.6061	-.5194	-.4119	
	B			-.3448	-.3510	-.5992		
↗				-.4721	-.7479			
				-.5389				n: 1

Chapter 9

Time Series of Disasters

Disasters in British coal mines between 1851 and 1962 provide a well studied data base, mainly analysed for the shape of the distribution of times between successive disasters. Here the series is treated as one of variations in the local rate of fatalities. Another series, also showing erratic fluctuations, sometimes due to unidentified exogenous factors, is that of the reported monthly sightings of UFOs in the USA over an 18-year period. Both these series raise interesting questions for social psychologists, yet can require quite different methods of analysis to explore their dynamics. There are big differences in temporal scale, and in the number of uncontrolled and, indeed, unidentified and uncontrollable variables, from the specifiable psychophysics of Chapter 1, but some of the considerations about identifying chaotic dynamics still surface in analogous ways.

Many real situations can generate series which are a mixture of chaotic, periodic and stochastic noise components. The actual time scale, in milliseconds, days, months, or years, is immaterial unless one also has more data for interpolation. If the series are long (over 2,000 points) then it may be possible to sample by coarsening the resolution and look for self-similarity and fractal properties, but this is not generally true. Nonstationarity of the nonlinear dynamics, overwritten by second-order stochastic noise, is to be expected in some psychophysical and psychophysiological data. It is necessary to break a long series into successive subseries to examine for nonstationarity, and this in itself invalidates the use of algorithms that depend strongly on delay coordinates in order to estimate the correlation dimension. It is to be expected that, even in stationary processes, the local largest Lyapunov exponent will fluctuate, approximately as a log normal distribution, and what is observed under nonstationarity is fluctuation from both sampling and dynamical causes. As Schaffer et al (1988) remarked, "current methods for estimating the largest Lyapunov exponent choke on nonuniformity", so that even if a series is long it is not necessarily a stable basis for estimation.

The method explored here uses the fluctuations of the local largest Lyapunov exponent, the entropic analogue of the Schwarzian derivative (ESf), the confidence intervals of the ESf derived from surrogates, higher order convolutions in the time domain, and the eigenvalues of the matrix of the ESf's of those convolutions. It has been found that this approach can be applied to series as short as 120 points; the present study now reduces that to 72 points (each set of 72 one-month points corresponding to a six-year period in the source data) to see what emerges as compared with previous analyses in the relevant literature.

If a process is nonstationary in its parameters for all but the most complicated hierarchical models, then its identification in the presence of stochastic noise is impeded, particularly if it is simple in its stationary epochs. However, if it is complex or chaotic, it can be more robust to the effects of perturbing noise (Schaffer, Ellner & Kot, 1986).

The Sugihara and May (1990) approach is essentially black box, its focus is on seeing if predictability is a sufficient way of distinguishing random from chaotic trajectories, under some assumptions about stationarity. This is still necessary, as the usual statistics, D2 and Lyapunov, are not helpful and, indeed, dimensionality itself is not the most informative property to compute for real but short biological series. As soon as a series shows dead or interpolated epochs then it is not stationary, but can be put into a finite-state Markov Chain if the dead or null epochs are themselves defined as occupying one state of the system. The problem then becomes identifying the stationarity of the state transition probability matrix in the Markov chain. Any such model needs tuning parameters (Robert & Casella, 1999); the role of the partitioning constant δ in ESf is just one such parameter.

Limits on Identifiability

The New York measles series was used by Sugihara and May (1990) as one test set to develop methods for discriminating between stochastic random processes and chaotic processes. It is well known that such discrimination cannot always be achieved. In fact, the confusability has a practical application. Robert and Casella (1999, Chapter 2) review a number of uniform pseudo-random number generators which reproduce the behaviour of an i.i.d. sample of uniform random variables when compared through a usual set of tests. Some of these generators are mathematically chaotic attractors, which have such a long periodicity that recurrence of identical subseries is not encountered in practical applications; that is, there are chaotic series that for a specific sample of tests will pass as random. Proceeding in the opposite direction, there may be series which are essentially random by some stochastic definition but will mimic a chaotic trajectory.

The properties of chaotic series which may be tested for as almost definitional include sensitivity to initial conditions, local fractal dimensionality, positivity of the largest Lyapunov exponent, local predictability, and entropic functions. If some function of the system's control parameters is experimentally accessible, then it may be possible to push the dynamics through a sequence of bifurcations, revealing Feigenbaum dynamics, which could not be done in a stochastic process. The presence of noise and of nonstationarity or singularities will weaken the effectiveness of such tests; for example, the embedding dimension of an attractor may be a fixed integer, but its non-integer dimensionality on the invariant manifold could fluctuate. The orbits of the Lorenz equation (Davies, 1999, p.9) are a well known and frequently depicted case: "the generating point makes one or two circuits around one of the wings before switching to the other."

Each wing could be locally contained in two dimensions, but the wings are not coplanar and it needs more dimensions to hold the whole orbits. The Shilnikov attractor is a more extreme example; it runs to the centre of a spiral in a flat plane and then jumps out orthogonally and re-enters the spiral at its outer limit to begin the inward path again.

The frequency spectrum, and hence the autocorrelation spectrum, of a chaotic process is wide with some local peaks. But a mixture of white or coloured noise with some limited periodicity superimposed can create the same picture. Such mixtures may have a small positive largest Lyapunov exponent. Fourier analyses may be misleading; Davies (1999, p. 56) shows the Fourier amplitudes of the logistic map, where there is a strong period 4 component, but the orbit is not periodic.

Time series which have discontinuities in them or are suspected of having some sort of instability in their generation have been the subject of a diversity of statistical methods to explore where the discontinuities are located, whether or not they are abrupt, and what are the best representations of the discrete segments in the series lying between the discontinuities (Spall, 1988). In real data, there may be critical events, such as a heart attack, or a civil war, which trigger a shift in dynamics. The series may recover from interruptions or may be permanently changed or terminated. If the series is not isolated but has collateral variables running in their own series then the interaction between two or more parallel series may furnish valuable clues about where and what disruptions have occurred.

An elegant treatment of the various commoner sorts of discontinuities in univariate time series was created by Gordon and Smith (in Spall, 1988, chapter 14) using state space equations. It was restricted to change point identification and modelling in processes in which the evolution was linear between change points and the series was one of equal interval observations, but is still valuable because it models, within the same equations, steady-state, changes in level, change in slope, and transient outliers. However, it says nothing about higher-order dynamics, and could not be used to predict the location of the next irregularity. The series to be considered here could be partly modelled by state-space equations if we were only interested in identifying some discontinuities as departures from a hypothetical steady state. What exactly sufficiently defines a steady state is somewhat opaque as soon as consideration of higher-order dynamics comes into the picture.

There were many fatal accidents in the British coal mining industry between 1851 and 1963. The table used here from Andrews and Herzberg (1985) has a lower bound on the number of fatalities in one accident set at 10, and labels the series as one of 191 disasters. The earlier analyses by Maguire, Pearson and Wynn (1952,1953) are based on an incomplete series, and treat accidents as a point process, ignoring the actual number of deaths involved. The focus was on comparing methods to test the non-randomness of the distribution of intervals between accidents, which if random may approximate to a Poisson distribution, or to a Laplace transform. One might think that such an approach is unfeeling and irresponsible but, in fairness to those authors, they remark (p. 179): "A thorough study of time intervals between explosions in mines would take into consideration the number of men killed and the number at risk." They also commented that, "The problems of accidents are more human and more complicated than those of quality control. It is usually more important to extract the maximum amount of information from industrial accident data." Jarrett (1979) augmented the series and some internal nonuniformity in the rate of the process has been noted; it appears to change after 125 disasters (Cox & Lewis,1966).

The series is thus long enough to be partitioned into 64, 64, and 63 successive events, to apply the same methodology as in other series. If the series does, in fact, show a change in its generation process after 125 events, then the last subseries should be different from the first two. There were technological and social changes during this long period, but there also seem to have been periods of stagnation, as analysed by economists, during the late-19th century. It may be of interest that the series starts in 1851, but in 1852 the engineering employers locked out their men when they refused to accept systematic overtime (May, 1987, p. 236). Labour relations were bad and labour laws were significantly loaded against workers for much of the 19th century; legal recognition of trade unions did not occur until 1871. Mechanical extraction of coal from seams did not come into use in British mines until after 1900, at which date only 2 per cent of coal was mechanically extracted.

The Deaths Series, x

The great disaster of October 1913, with over 400 deaths, stands out. Its location is even more marked graphically if the series variable is squared. As the series does not use accidents in which fewer than 10 persons were killed, and we might reasonably expect that accidents involving one or two workers were much commoner than ones with over 10 deaths, the series has been, in effect, low-pass filtered, so estimates of the frequency distribution of both this series (x_j) and of the time intervals series (y_j) would need to be reanalysed with various cut-offs of higher values than 10 if independence of successive disasters is to be examined. The total labour force in mining, mostly coal, increased from about 200,000 in 1841 to 216,000 in 1851, 496,000 in 1881, 807,000 in 1901 and to 1,128,000 in 1913 (Crouzet, 1982, pp. 68, 268). After a peak of 1.2 million in 1920, the total labour force fell steadily to 702,000 in 1938 (May, 1987, p.329). The size of the population at risk is therefore not constant, nor are the working conditions in terms of size of mine or accessibility of the coal-faces.

For this series LLE(x) = .0366

Figure 9.1: The series of 191 accidents from 1851 to 1963, x

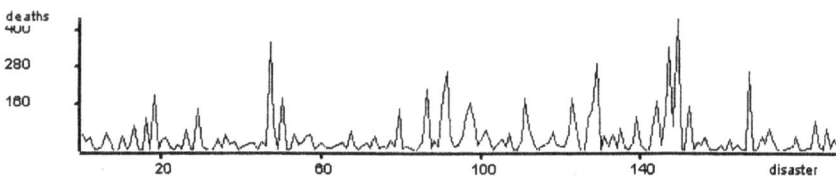

Figure 9.2: The same series, squared, x^2

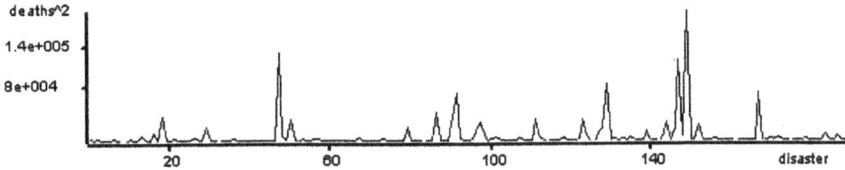

The Time Intervals Series, y

It is this series which has been the greatest focus of statistical theorising. Tests for its randomness have taken various forms. The graphical effect of squaring the variable is even more marked here, emphasising the contrast in the last subseries. This is matched in the table of descriptive statistics, in the first two moments. It is not matched in the first-order ESf values, precisely because those do not use the absolute values (and hence the first two moments) of the variables. Each time interval is the elapsed time in days since the previous disaster before the current one. There is, therefore, the other time series, in which the time interval is that immediately following a disaster before the next one occurs. One leads and the other lags on time when truncated cross-correlations between the two series x_j, y_j and $x_j, y_{(j+1)}$ are considered. LLE(y) = .0683

Figure 9.3: The time intervals series of 191 disasters, y

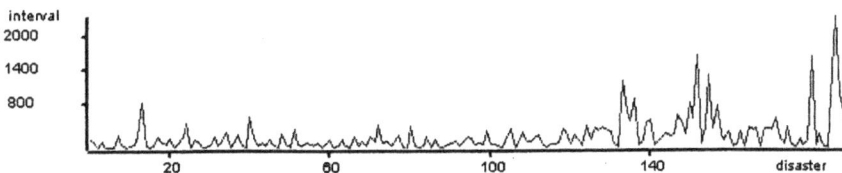

Figure 9.4: The same series, squared, y^2

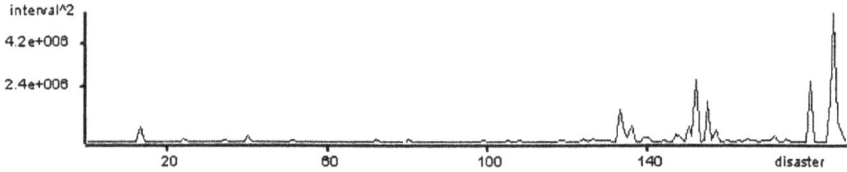

The Local Death Rates Series, z

Two series of local death rates are created by dividing the deaths x_j at disaster j by the time since the previous disaster y_j to give $_gz_j = x_j/y_j$; that is, deaths lag on time, or, for deaths leading time $_dz_j = x_j/y_{(j+1)}$, provided that if two disasters arise on the same day (this happened once in the long record) we set $y_j = 1$. These two z_j series may be explored by precisely the same methods as the first two series, but their interpretation demands great care. Why this is so may be seen by looking at some collateral statistics from other sources. LLE($_gz$) = .0366, LLE($_dz$) = .0745. It should be noted that these z series are the simplest sort of lagged cross-correlation spectra that we can compute.

Redmayne (1945, pp. 41-42) lists some data on accident rates in coal mines, with warnings that the data are not simply interpretable: "Mining is the most dangerous of all occupations, " and "the highest rate of accident mortality in mines is that in respect of the conveyors of material to the shaft". Obviously a large disaster will kill more than just this subgroup of the labour within one mine. Redmayne was aware that a single large disaster could bias the data for a time period, and hence calculated accident rates only over decennia, and also considered the number of shifts per worker employed, as this is variable and determines an exposure rate to risk.

Figure 9.5: The $_g z_j$ series;
time intervals lead the number of deaths in a disaster.

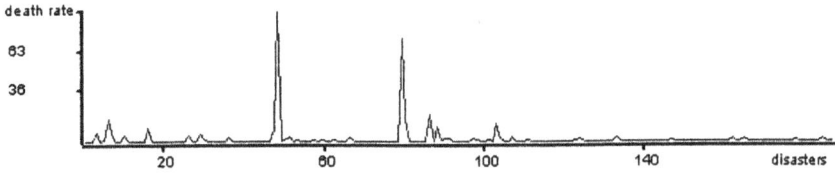

Figure 9.6: The $_d z_j$ series
time intervals lag the number of deaths in a disaster.

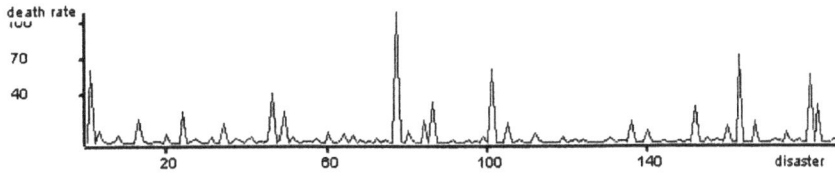

Table 9.1
Death Rate from Accidents, per 1000 employed
From Redmayne (1945), for 1873 to 1943

Decade	From	to	Rate
1	1873	1882	2.24
2	1883	1892	1.81
3	1893	1902	1.39
4	1903	1912	1.15
5	1913	1922	1.15
6	1923	1932	1.05
7	1933	1942	1.14
-	1943	only	1.00

Table 9.2
Death rates per 100,000 man-shifts worked,
surface and underground, from 1938

Year	Rate
1938	0.41
1939	0.37
1940	0.43
1941	0.46
1942	0.43
1943	0.36

It is perhaps better that we do not dwell on Redmayne's theories (1945, p. 43) about the psychology of the miner, "the Character and Ideocracy [sic] of the British Coal Miner", and what he calls racial differences between Celts and Danes. In interpreting the figures for 1913 to 1922, he eventually notes that the industry was under complete government control during 1917 to 1921, and "there were then strikes and unrest greater that at almost any other period" (no figures given). It would have been helpful if the relations between strike rates and death rates were tabled; it is a bit harder to get killed if one is out on strike. The whole text reads like a sustained apologia for private capitalism, it was written and published at a time when nationalisation of the coal industry was under active consideration. Of course, that came to pass. The relevance of Redmayne's tables for our current purpose is to display the changes in the generation of data during the total period under consideration.

When interpreting the number of deaths on one disaster, which means from one mine, one needs to know how big the mine was. Crouzet (1982, p. 270) notes that, "in 1913 there were 3289 collieries worked by 1589 separate firms. The average colliery had a workforce of 340 ... more than one third of the collieries were worked by fewer than 50 miners. The great majority employed from 50 to 2,500 without their size having any influence on their efficiency." If 35 miners were killed in one disaster, it means something quite different if the mine employed 50 men, rather than employing 500 or 5,000.

Having got the four series x, y, $_g z$, $_d z$ their descriptive statistics for each subrange may be tabulated.

Table 9.3
Descriptive Statistics including ESf on the four series

Statistic	x_1	x_2	x_3	y_1	y_2	y_3
mean	41.42	53.85	58.13	113.14	133.12	395.92
s.d.	55.23	55.38	83.87	137.76	115.98	464.57
variance	3050.21	3067.27	7035.17	1.8×10^5	1.3×10^5	2×10^6
Skewness	3.6026	1.9181	2.7049	2.8234	.8487	2.0776
Kurtosis	15.9097	3.2257	7.5542	10.3115	-.4055	4.6043
Slope	0.1620	0.3448	-1.1566	-0.4831	1.6635	1.5907
ESf	-.3769	-.2322	-.1872	-.4747	-.4739	-.4301
-	$_g z_1$	$_g z_2$	$_g z_3$	$_d z_1$	$_d z_2$	$_d z_3$
mean	2.86	2.71	0.46	4.79	5.34	5.16
s.d.	11.53	9.44	0.75	10.38	15.98	12.63
variance	132.87	89.13	0.55	107.64	255.47	159.62
Skewness	7.0063	6.1845	2.5981	3.6121	5.0138	3.7720
Kurtosis	50.4346	41.7792	6.8722	14.2947	27.1255	14.9990
Slope	.0407	-.0698	-.0019	-.0589	-.0993	.0979
ESf	-.2158	-.6152	-1.0773	-.8236	-.2161	-.2509

The series $_g z_j$ is one in which it is assumed that an accumulator process builds up the probability of an accident over time until it is eventually triggered, its severity being proportional to the time for risks to accumulate, whereas the $_d z_j$ series assumes that an accident exhausts the potential for another accident as a function of its own severity, so that the delay to the next accident is a proportional function of what has just occurred. It is quite possible for the process as a whole to shift from $_g z$ to $_d z$ (or the reverse) during its evolution; this is a form of dynamical nonstationarity.

The higher-order ESf matrices can also be computed on these subseries.

Table 9.4
Higher-order ESf matrices for the subseries of deaths; x
Italicised values are within the 95% c.i. for random surrogates

Block	ESf	b(1,2)	b(1,3)	b(1,4)	b(1,5)	b(1,6)
1(of 64)	-.3769	.1633	*.0643*	-.2771	*.1338*	-.0820
		.4008	.1329	*.2307*	*-.1311*	
		.2008	.2601	.2327		
		.2096	-.1919			
		-.6567				n: 5
2(of 64)	-.2322	*.0901*	-.2261	-.2299	.3409	.1961
		-.0581	*-.1435*	-.1124	.2398	
		-.2906	-.3552	-.0167		
		-.1376	*-.1484*			
		.2420				n: 3
3(of 63)	-.1872	.3653	.2219	.2623	*.2442*	.1436
		.2610	*0955*	.1412	*.1504*	
		.2219	*.2569*	*.0494*		
		-.2392	.2113			
		.2219				n: 6

Again, the interest is in evidence of heterogeneity between the three sub-series. The ESf values for the first-order series are reduplicated for cross-checking with the previous table of descriptive statistics. From previous analyses of time series in this fashion, it can be read that these series are a mix of random and non-random higher-order relations within the sequential dynamics, the deaths series is not one which is locally predictable with any confidence. The deviant subseries in ESf terms appears to be the middle one, not the last one, as suggested by the earlier analyses of time intervals by Maguire, Pearson and Wynn (1952). The middle period is the least random in terms of higher-order dynamics, for both the deaths and the intervals subseries. A few very disastrous accidents within a series of much smaller ones can induce odd sequential dynamics.

Discussion

A classical problem in data analysis is that of inferring causality from a correlation; the inference is usually invalid if unsupported by contextual knowledge. In time series, there are additional possibilities for inferring the causal direction of links between variables. The additional information in time series data is that which gives some insight into the temporal direction of effects (as, for example, in path analysis); if the cross-correlations in one direction at one or more lags are very different from those in the opposite direction, then some sorts of causality are supported and others ruled out. The extension from more familiar ideas here involves both the $_gz$ and $_dz$ series constructed from the two x and y data series, and the ESf at first-order and higher-order values; for example, the ratios of the first-order ESf for $_gz$ and $_dz$ are 0.262, 2.847, and 4.294 for respectively subseries 1,2,3, whereas for the ratios x/y they are 0.794, 0.490 and 0.435.

It is usual to compute the cross-correlations between two parallel series and see if one lags or leads upon the other. This interaction between two series may itself be nonstationary and what lags at one place may lead at another. If such changes are happening, then reduction of a total unbroken multivariate time series to one based on the Jacobians of the cross-correlations at each lag can be seriously misleading (Gregson, 1983, Chapter 6).

The four tables of higher-order ESf matrices differ from one another. Changes in one do not match changes in another but, considering the complicated causality of the collection of coal mines spread over regions with different geology, sizes, technology, ownership, and apparently labour-management relations, there is an underlying causal heterogeneity that is not present to anything like the same degree in the measles series. Hence any deductions are generalisations which at best will show differences between different time periods.

The y series of time intervals between accidents is, indeed, nearest to a Poisson distribution if one allows the ratio of mean/s.d. to be the indicator and the large change does then show in the last subseries where accidents become rarer in time. There is not much difference in the average number of deaths (the x) series per accident over the the three subseries, but obviously if one computes deaths per time (in months, and perhaps corrected for the size of the labour force as it increased up to 1913) then conditions improved in one respect.

Table 9.5

Higher-order ESF matrices for the subseries of intervals; y

Italicised values are within the 95% c.i. for random surrogates

Block	ESf	b(1,2)	b(1,3)	b(1,4)	b(1,5	b(1,6)
1(of 64)	-.4747	-.0192	-.2556	-.1251	-.0468	-.2371
		.0144	-.1089	-.2369	.2223	
		-.0206	-.4565	-.0659		
		-.0284	-.3653			
		-.2859				n: 6
2(of 64)	-.4739	-.0248	-.4247	-.4662	-.3412	.1588
		.1409	-.0989	-.2655	.0126	
		-.3062	-.1773	-.1182		
		-.1015	-.3507			
		.1272				n: 2
3(of 63)	-.4301	.2412	.0117	-.0784	-.1403	.4620
		-.1386	-.0261	.0804	-.2372	
		.2940	-.0690	-.0859		
		-.2283	.0285			
		.0405				n: 3

The four higher-order ESf tables suggest that the x series is noisy throughout. The y series becomes much less noisy as it evolves, mainly due to the fact that relatively little happens after about 1920, as technology and prevailing economics change. This is also is reflected in the $_g z_j$ series, but not in the $_d z_j$ series, which is less noisy. There is thus stronger evidence for an accumulator process in which the time intervals are longer after serious accidents. Accidents are not so much waiting to happen as many secondary factors accumulate (as in a Poisson process) but rather that an accident induces, for a while, more caution in all pits as knowledge of the last disaster is diffused or knocks out, for a while, from the pool of pits at risk those with larger current accident potential. This distinction is completely lost if only the y series is analysed, as the earlier work of Maguire, Pearson and Wynn (1953) employed.

Table 9.6

Higher-order ESf matrices for the death rates series; z

This is $_gz_j$ where deaths lag on time intervals

Italicised values are within the 95% c.i. for random surrogates

Block	ESf	b(1,2)	b(1,3)	b(1,4)	b(1,5)	b(1,6)
1(of 64)	-.2158	-.1136	*.0601*	.1507	.1735	-.1760
		-.2762	*.0145*	*.0756*	.1518	
		.2824	.2088	*.2007*		
		-.1976	.0429			
		.1705				n: 5
2(of 64)	-.6152	-.7049	.2126	.2610	-.1863	.2610
		.4008	-.0978	*.1134*	*.1102*	
		.4008	.0078	.2816		
		-.0434	*.1344*			
		.1879				n: 5
3(of 63)	-1.0773	.1610	*.4008*	.0247	.0651	.4008
		.2137	.0382	*.2610*	.2108	
		.0532	.2990	.1800		
		-.0182	.0054			
		.3044				n: 3

Table 9.7
Higher-order ESf matrices for the death rates series; z
This is $_d z_j$ where deaths lead on time intervals
Italicised values are within the 95% c.i. for random surrogates

Block	ESf	b(1,2)	b(1,3)	b(1,4)	b(1,5)	b(1,6)
1(of 64)	-.8236	-.3556	.3613	-.4269	.3773	.2498
		.1789	.1714	.2946	.2813	
		.2345	-.1805	.2240		
		.1565	.3130			
		.1134				n: 3
2(of 64)	-.2161	.1639	.3337	.4163	.1728	.4281
		.2345	.0875	.2126	-.4886	
		.2656	.2219	.4581		
		-.2253	-.0156			
		-.1453				n: 4
3(of 63)	-.2509	.0874	-.4146	.1602	-.2999	.0386
		-.1616	.2705	.5082	.3927	
		-.1589	.3374	.1295		
		-.0426	.2946			
		.1820				n: 3

Series of Irrational Beliefs

The series of coal mine disasters has some real identifiable bases in physical causality, as well as in the bitter social confrontations between miners and mine owners. We can, in part, identify some factors as changes in the technology of mining and even in the medical ways in which the injured are saved from death after being rescued; such factors changed over the 19th and 20th centuries. But there are series of events in which scientific or technological explanations are lacking or perhaps at least contentious. One example is the series of UFO (or, popularly, "flying saucers") sightings in the USA, recorded as monthly totals (Condon, 1969, Section V, Table 1).

The series is not random and shows local outbursts. In this respect, it superficially resembles series of volcanic eruptions, epidemics such as the Black Death in Europe, or fanatic attacks on witches in the 17th and 18th centuries. For volcanoes, we now know much of the geophysics, though we cannot predict precisely well in advance a particular eruption. Epidemics such as measles have periodicities, but some psychological series are aperiodic and based on irrational beliefs that seem to disperse something like biological epidemics. It is tempting to see if irrational series are chaotic. There is little point in testing them against random surrogate series because we know enough about their social causality to know they are not random. It is not informative simply to be told again that they have some underlying partially determinate causality.

Figure 9.7: The raw data of UFO sighting frequenciess
Eighteen years from 1950, as reported month by month

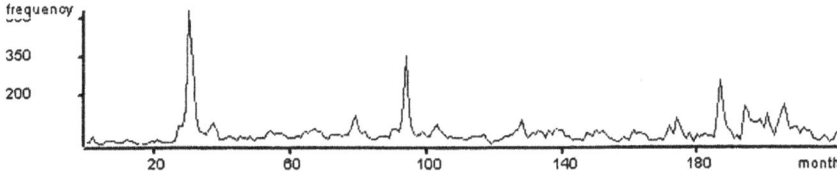

The full data are given in Table 9.9. It is not clear if multiple sightings by one individual are included, or if simultaneous sightings by a group of observers counts as one and, obviously, the return rate of observations is unspecified; that is, the data are properly sightings reported to the study and the rate of reporting is itself a hidden variable. The relation between sightings and reporting sightings can be a variable that is a function of other external variables, such as media feature programs and fictional reports of aliens invading the USA. The aspect of interest are, however, the non-stationarity, the local peaks and the autocorrelation spectra. The autocorrelations vary markedly with the transformations of the raw frequencies, and are maximised under a reciprocal transformation. There is no suggestion of an annual periodicity. Showing where and how the non-stationarity arises demands that the series is explored under other transformations and using filters in the sense of Chapter 1; a low pass filter merely repeats the pattern of Figure 9.7 with smoothing the peaks, a high-pass filter is now much more informative. This is shown in Figures 9.8 and 9.9. Calculating indices of chaoticity on filtered series is not commonly done, but we note that Abarbanel (1996, page 93) anticipates this when he writes, "we are seeing data with small scale motions suppressed yet with very interesting dynamics How the Lyapunov exponents of a system vary with spatial averaging is the issue."

If we wish to use symbolic dynamics (Gregson, 2005), then the UFO sightings series has to be coarse scaled by partitioning into a closed set of segments and labelling the segments each with a dummy variable. We will use 5 segments and then put the data into a 5-state Markov transition probability matrix. The simplest partitioning is into equal width ranges, a maximum a priori entropy assumption. This is in fact false, the low values as much commoner, so three solutions are presented.

Figure 9.8: Reciprocals of monthly UFO sighting frequencies
derived From Figure 9.7

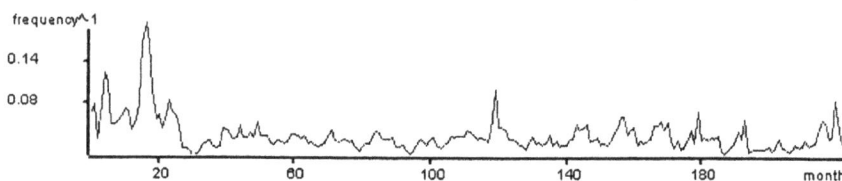

Table 9.12 (see also Figure 9.7) is based on using the reciprocals of the raw frequencies, which thus expresses the process in terms of local rates of sightings. This also gives us a long term prediction of the frequency and rate distributions, coarse scaled, of UFO sightings after 1968, under assumptions of stationarity dynamics.

The series is long enough to get some estimates of the largest Lya-

Figure 9.9: Highpass filtering of UFO sighting frequencies
derived From Figure 9.7

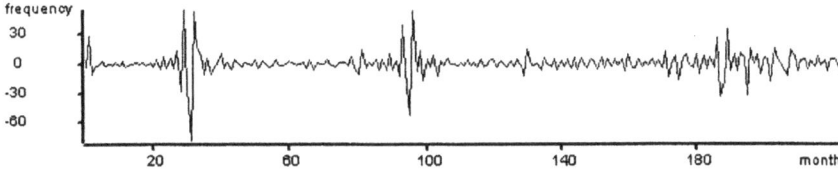

punov exponent (LLE). In all cases it is positive but very small, suggesting edge-of-chaos; For raw frequency data, LLE = +.0394, for a $\log_{10}$ transform, +.0862, for reciprocals, +.0335, for high pass filtering, +.0478, and for low pass filtering +.0643.

ESf are computable, for the full series ESf =-.5745 with $\delta = .005$, for the first 9 years,-.5009, and for the last 9 years -.4810. This is compatible with the appearance of non-stationarity.

Table 9.8 Autocorrelation Spectra under Transformations
The raw frequencies, their logs and their reciprocals are tabled in parallel.

lag	autocorrel	$\log_{10}$ trans	reciprocal
1	.5938	.7418	.7611
2	.2640	.5392	.5357
3	.1481	.4220	.4119
4	.0672	.3177	.3501
5	.0222	.2305	.3156
6	.0197	.1915	.3103
7	.0585	.1751	.2922
8	.0436	.1922	.2768
9	.0014	.1718	.2543
10	-.0139	.1737	.2662

So the partitioning points are 0,.2,.4,.6,.8,1.0 for Table 9.8, and (quite arbitrarily and heuristically) 0,.1,.2,.6,.9,1.0 for Table 9.9. The reciprocals have a skewed distribution so 0,.02,.04,.08,.2,1.0 have been used.

The stationary state vectors in Tables 9.10, 9.11, and 9.12 are based on observed convergence after 20 iterations. The most interesting and potentially informative are those from the reciprocals of the frequencies in Table 9.12.

Table 9.9: Monthly Returns on UFO sightings

	Jan	Feb	Mar	Apr	May	Jun	Jul	Aug	Sep	Oct	Nov	Dec
1950	15	13	41	17	8	9	21	21	19	17	14	15
1951	25	18	13	6	5	6	10	18	16	24	16	12
1952	15	17	23	82	79	148	536	326	124	61	50	42
1953	67	91	70	24	25	32	41	35	22	37	35	29
1954	36	20	34	34	34	51	60	43	48	51	46	30
1955	30	34	41	33	54	48	63	68	57	55	32	25
1956	43	46	44	39	46	43	72	123	71	53	56	34
1957	27	29	39	39	39	35	70	70	59	103	361	136
1958	61	41	47	57	40	36	63	84	65	53	33	37
1959	34	33	34	26	29	34	40	37	40	47	26	10
1960	23	23	25	39	40	44	59	60	106	54	33	51
1961	47	61	49	31	60	45	71	63	62	41	40	21
1962	26	24	21	48	44	36	65	52	57	44	34	23
1963	17	17	30	26	23	64	43	52	43	39	22	22
1964	19	26	20	43	83	42	110	85	41	26	51	15
1965	45	35	43	36	41	33	135	262	104	70	55	28
1966	38	18	158	143	99	92	93	104	67	126	82	40
1967	81	115	165	112	63	77	75	44	69	58	54	24

Mitchener and Nowak (2004) found a transition matrix for a process which they considered resembling a Shilnikov attractor trajectory; in this 5-state representation, the upper left 3×3 submatrix is one part of the system, which is fully connected, the lower right 2×2 high amplitude part represents a subregion into which the trajectory can transiently jump. These dynamics are chaotic but, in a short realisation, appear non-stationary. We return to this topic in the next chapter. Another way of visualising the system is to think of the 3×3 submatrix as partitioning the system's dynamics and the transient jumps to constitute themselves a slow series with aperiodic intervals between the jumps.

Table 9.10: 5-state transition matrix with equal partitioning
The last columns are the respective stationary state vectors

.967	.029	.005	.000	.000	.9330
.545	.273	.000	.182	.000	.0491
.000	1.000	.000	.000	.00	.0045
.500	.000	.000	.000	.500	.0089
.000	1.000	.000	.000	.000	.0045

Table 9.11: 5-state transition matrix with skewed partitioning

.884	.116	.000	.000	.000	.7319
.378	.467	.156	.000	.000	.2010
.167	.333	.333	.167	.000	.0537
.000	.500	.000	.000	.500	.0090
.000	.000	1.000	.000	.000	.0045

Table 9.12: 5-state transition matrix for reciprocal values

.671	.278	.051	.000	.000	.3491
.247	.613	.140	.000	.000	.4080
.070	.256	.558	.116	.000	.2009
.000	.250	.375	.250	.125	.0374
.000	.000	.000	1.000	.000	.0047

In Table 9.12, the meaning of fast and slow submatrices is reversed: the upper 3×3 submatrix is now associated with slow dynamics, and the lower right 2×2 submatrix with faster changes. Because of the social changes in UFO beliefs it is very doubtful if the stationary state vectors have any long-term validity; possibly they could be replaced by beliefs in terrorism, as another form of social near-panic. Even more transitory were the millennial vigils around the year 2000 (Gregson and Gregson,1999).

An argument might be advanced for treating series such as the UFO sightings as convolutions of a least two other processes: a social inertia or contaminative effects within a social group, and perturbations from critical incidents in the cultural environment. That would lead us back into parallels with the spread of infectious diseases about which many stochastic models have been developed; some examples were noted in the last part of Chapter 1.

Note

The coal mine disasters series are taken from Andrews and Herzberg (1985).

The UFO sighting frequencies are taken from a US Air Force commissioned report by Condon (1969).

Chapter 10

Perron-Frobenius at the Edge of Chaos

From the approach of symbolic dynamics, any psychophysiological time series may be given a square non-negative matrix representation that is then treated as the generator of a Markov chain. This has eigenvalues that, if the matrix is scrambled, that is effectively not degenerate, give a picture of the complexity of the dynamics. That picture is computed for two time series: one theoretical and homogeneous, resembling a Shilnikov attractor, and the other from real physiological data that are very unstable with transient outliers. A comparison is made with indices of entropy and chaos for each of 10 data sub-blocks. No index in itself provides a satisfactory representation of the total dynamics, but the differences between the indices are intrinsically informative. Assumptions of linearity are universally invalidated. The use of the entropic analogue of the Schwarzian derivative (ESf) leads naturally into the calculation of Kullback-Leibler information measures as asymmetric proximity indices between subseries of the data. The full matrix of these indices has eigenvalues that are informative concerning the non-stationarity of the process. The matrices that we create for a representation of psychophysiological and psychophysical time series are Markovian, and are necessarily square and non-negative. They may also be sparsely filled, and quasi-cyclic. We know both from the fundamental mathematics, and from examples constructed (Mitchener and Nowak, 2004) that, if we accept the positivity of the largest Lyapunov exponents as a sufficient indication of chaos, of one sort, then the processes being represented in transition probability matrix form may be chaotic. Possibly the most important extension of these ideas is to non-homogeneous Markov chains as a structure for non-stationary psychophysics.

As I will want to examine some real and theoretical Markovian matrices and their associated eigenvalues, it is proper to begin by restating the Perron-Frobenius theorem (Frobenius, 1912, Seneta, 1973) for primitive matrices: Suppose that T is an $n \times n$ non-negative primitive matrix. Then there exists an eigenvalue r such that:

(a) r real, > 0;

(b) with r can be associated strictly positive left and right eigenvectors;

(c) $r > |\lambda|$ for any eigenvalue $\lambda \neq r$;

(d) the eigenvalues associated with e are unique to constant multiples;

(e) if $0 \leq B \leq T$ and β is an eigenvalue of B, then $|\beta| \leq r$, and $|\beta| = r$ impies $B = T$.

(f) r r is a root of the characteristic equation of T.

There are other statements that hold from this theorem (Ding and Fay, 2005) which find application in the estimation of geometrical limits in n-space.

I am also going to need to make reference to the well-known scrambling property of Markov matrices, namely
An $n \times n$ stochastic matrix $p + \{p_{ij}\}$ is called a *scrambling* matrix, if given any two rows α and β there is at least one column, γ, such that $p_{\alpha\gamma} > 0$ and $p_{\beta\gamma} > 0$.

A corollary follows: if $Q = \{q_{ij}\}$ is another stochastic matrix, then for any Q, QP is scrambling, for fixed scrambling P.

A Leaky Bistable Matrix

In a study of language shifts and chaotic dynamics, Mitchener and Nowak (2004) created a scrambled matrix. The matrix, call it LS, has one free parameter $0 < \mu < 1$, and is

$$LS = \begin{pmatrix} 0.75 & 0.2 & 0.01 & 0.04 & 0 \\ 0.01 & 0.75 & 0.2 & 0.04 & 0 \\ 0.2 & 0.01 & 0.75 & 0.04 & 0 \\ 0 & 0 & 0 & \mu & 1-\mu \\ 1-\mu & 0 & 0 & 0 & \mu \end{pmatrix}$$

The authors investigated the results of varying μ over the closed range (.725, .760), in terms of the properties of trajectories generated by the recursive application of LS. They claimed that the dynamics can resemble a Shilnikov attractor (1966), and are critical around $\mu = .735$. The eigenvalues for LS with $\mu = .725$ are

$$.978, .716 \pm .069i, .645 \pm .165i$$

and .978 is the *spectral radius*.

This resemblance is qualitative in terms of the shape of trajectories but is not strictly established in terms of a mathematical identity. Let us call this matrix bistable, as it involves two connected submatrices, within each the process can meander for a while in a cyclic fashion, and leaky because there exists a non-zero probability of jumping between one submatrix and the other in both directions. If it were not leaky, then the structure would break into two separate autonomous cycles. We examine some of its generic properties; the questions raised here can also be treated as special problems in symbolic dynamics, but to do that requires a more advanced mathematical treatment (Blanchard, Maas and Noguiera, 2000).

Probability partitioning

The probability of making a transition between states $j \to k$ is independent of the random input p_t and strictly proportional to the elements p_{jk}.

The row sums of p_{jk} in LS are all unity, by definition. So take the unit interval $(> 0, 1)$ within which the random variable p_t lies, and partition it in line segments s_1, s_3, s_3 proportionately to the defined values of p_{jk} in S_1, S_2, S_3, and similarly for $\mu, 1 - \mu$ in S_4, S_5. Then if and only if p_t lies in segment s_{jk}, given S_j, the resulting transition is $j \rightarrow k$. This convention preserves the Markov structure of LS.

An estimate of the stationary state vector from a generated sample series is

$$.078, .402, .273, .134, .111$$

Table 10.1: Descriptive statistics of a sample LS series

Block	LLE	mean	Kurt	ESf	D2	H	ApEn
1-180	+.051	2.79	-0.353	-1.2541	1.716	.2837	0.535

Key: Block: iterations. LLE: largest Lyapunov exponent. mean: average coded state. Kurt: kurtosis. ESf: entropic analogue of Schwarzian derivative (Gregson, 2002). D2: fractal dimensionality. H: Hurst index. ApEn: approximate entropy ($k = 10, \delta = .005$).

A Tremor Series

A series with a very unstable appearance and irregular transient extreme deviations is s14r45of.d from the PhysioNet archives. Only the first 4000 steps are used, in 10 blocks of 400. The dynamics are, on visual inspection, nonlinear and non-stationary. Deviations can be isolated in time or come in bursts. Every estimate of the largest Lyapunov exponent has been positive but, in some cases, marginal.

These series are wildly fluctuating and their irregularity is reflected in most of the computed indices. It is recognised that complexity changes in time series as detected by information (entropy) measures (Tores and Gamereo, 2000). The distribution of the dependent variable (as in the vertical axes in the graphs) is nothing like Gaussian, and would not he readily represented in the frequency domain, as Fourier analyses cannot han-

Figures 10.1, 10.2, 10.3: Blocks of the Tremor Series

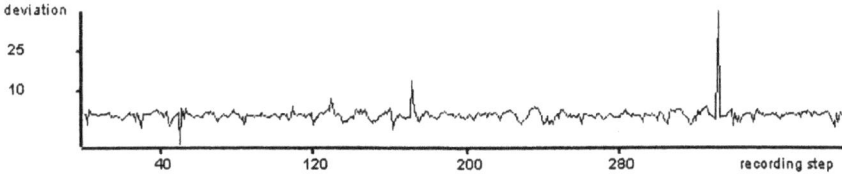

Tremor series, first block 1-400

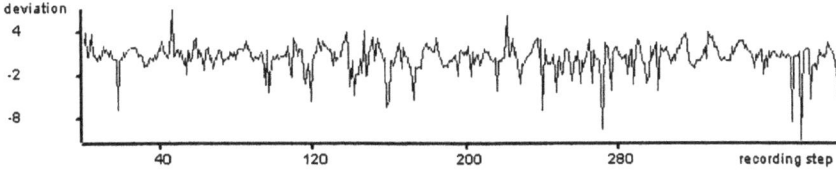

Tremor series, block 3, 801-1200

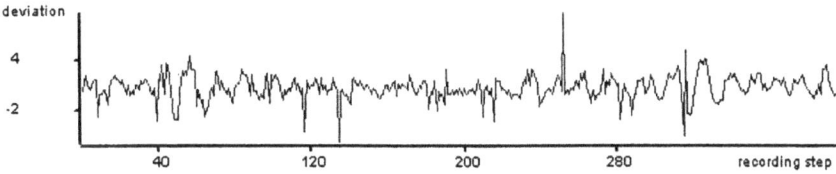

Tremor series, block 9, 3201-3600

dle local aperiodic spikes. Autocorrelations would be similarly misleading except as long-term averages. It is possible to treat this sort of series in a Markov chain filtering. It is possible that the leaky bistable matrices introduced by Mitchener and Nowak (2004) are a candidate model. There appear to be two levels of tremors, low amplitude high frequency and intermittent aperiodic high amplitude, which co-exist. High-pass and low-pass filtering could be used to partial these out.

Table 10.2: Descriptive statistics of Parkinsonian Tremor series

Block	LLE	mean	Kurt	ESf	D2	H	ApEn
1-400	+.034	.812	133.41	-.3897	1.835	.1647	1.112
401-800	+.078	.717	19.63	-.5572	1.806	.1928	1.122
801-1200	+.083	.529	6.43	-.5027	1.806	.1937	1.120
1201-1600	+.052	.725	71.66	-.5578	1.780	.2200	1.188
1601-2000	+.009	1.033	156.22	-.5025	1.932	.0684	1.092
2001-2400	+.073	.831	5.93	-.6258	1.788	.2118	1.245
2401-2800	+.096	.640	5.82	-.6642	1.790	.2103	1.250
2801-3200	+.065	.937	23.80	-.6354	1.780	.2201	1.259
3201-3600	+.080	.756	5.17	-.6146	1.836	.1641	1.219
3601-4000	+.023	1.106	122.78	-.4891	1.750	.2498	1.111
−	−	−	−	−	−	−	−
cftv	.449	.207	1.045	.145	.298	.250	.054
regr	1	-4.94	-1874.52	-1.89	-5.08	0.78	1.44
corr	1	.825	.910	.660	.289	.461	.635

Key: Block: beats in sequence. LLE: largest Lyapunov exponent. mean: average i.b.i. Kurt: kurtosis. ESf: entropic analogue of Schwarzian derivative (Gregson, 2002). D2: fractal dimensionality. H: Hurst index. ApEn: approximate entropy ($k = 10$, $\delta = .005$). cftv: coefficient of variation ($\sigma/|\mu|$). regr: slope of regression of variable on LLE. corr: product-moment correlation of LLE with variable.

The ESf and LLE measures can have associated confidence limits or filtered estimates computed as well, giving more insight into the variability between successive blocks of data. Comparing Tables 10.2 and 10.3 we see immediately that all the ESf values are outside their random surrogate 95% c.i. range; none of the blocks are random i.i.d. series.

The reason for computing the LLE values after low or high pass filtering is to explore the possibility that the chaotic component is either in a slow carrier wave or a fast low-amplitude part; this will only work if the fast and slow parts are widely separated in their dynamics. In this series, both the slow and the fast components are associated with the transient aperiodic outlier spikes, suggesting that there are three dynamical sub-systems involved. There are still computational difficulties, as LLE values from short time series are liable to be biased (Rosenstein, Collins and De Luca, 1993), the blocks here are much shorter than what physicists like to call short.

Table 10.3: Further c.i. statistics of Table 10.2

Block	LpLLE	HpLLE	-rsESf	rsESf	+rsESf
1-400	+.069	+.055	-.167	-.146	-.125
401-800	+.077	+.103	-.280	-.269	-.258
801-1200	+.114	+.113	-.363	-.349	-.336
1201-1600	+.093	+.029	-.242	-.223	-.204
1601-2000	+.036	+.034	-.224	-.208	-.192
2001-2400	+.090	+.101	-.338	-.327	-.315
2401-2800	+.127	+.125	-.397	-.389	-.381
2801-3200	+.069	+.086	-.314	-.302	-.290
3201-3600	+.118	+.093	-.375	-.363	-.351
3601-4000	+.059	+.031	-.161	-.143	-.125

Key: LpLLE: LLE after low pass filter. HpLLE: LLE after high pass filter. -rsESf: lower 95% c.i. bound for random surrogate ESf. rsESf: mean random surrogate ESf. +rsESf: upper 95% c.i. bound for random surrogate ESf.

In Table 10.4 the spherical radius is the same as the largest eigenvalue, Egv 1. Note that the number of complex conjugate pairs in the roots is not necessarily the same as for the LS matrix, which has quite different and homogeneous dynamics. These real data are not stationary and hence are treated in short blocks to expose the breakdown of homogeneity. Note that the dynamics drift over the total series of 4,000 data points; in some epochs the outlying tremors seem to be predominantly in one direction, in others in the opposite one. In blocks 5 and 6 the effective rank of the matrix reduces to 4.

Table 10.4: Eigenvalues of Markov matrices of the Tremor Series

Block	Egv 1	Egv 2	Egv 3	Egv 4	Egv 5
1	1.000	.291	-.2	.041 + .066i	.041 - .066i
2	1.001	.316+.015i	.316-.015i	.021	.012
3	1.000	.283	-.063	.052+.024i	.052-.024i
4	.999	.381	.236	.103	-.005
5	.999	.259	-.056	-.002	$< .001$
6	1.044	.237	-.048	.028	$< .001$
7	.999	.314	.118	-.064	-.002
8	.999	.219+.046i	.219-.046i	.165	-.006
9	.999	.351	.234	.115	.068
10	1.004	.496	-.010 + .051i	-.010 - .051i	-.006

The values of Egv 1> 1 are computational errors, the largest eigenvalue of a row stochastic matrix is less than or equal to unity (Karpelevich, 1951). The stationary state vectors are the solution of the matrix equation

$$\mathbf{T}_\infty = M^\infty T_0$$

The state S5 refers to the largest positive transient tremors, expressed as deviations from a resting position. The solutions in Tables 10.4 and 10.5 have to be obtained by shifting, in some blocks, the partitions that create the state occupancies from the raw data sequences, as the tremors are very irregular in direction and amplitude as well as their inter-tremor intervals. In all cases, I have moved partitions, where necessary, to span the rescaled values and to ensure that all states are occupied. The transition probability matrices are then in all cases scrambled.

Information measures as non-stationarity indices

The Schwarzian derivative has a long history but its importance in nonlinear dynamics only emerged relatively recently (Gregson, 1988). Combining this derivative of C^3 diffeomorphisms with an entropy measure over local short time-series is an exploratory innovation which has some useful properties. We repeat some of the definitions here as a precursor to using the measure in Kullback-Leibler form.

Table 10.5: Stationary state vectors for Table 10.4.

Block	p(S1)	p(S2)	p(S3)	p(S4)	p(S5)
1	.0485	.0485	.3495	.5340	.0194
2	.0977	.2882	.4662	.1429	.0050
3	.0676	.0675	.3076	.5323	.0251
4	.0201	.1378	.7569	.0827	.0025
5	.0301	.8346	.1228	.0100	.0025
6	.0102	.0050	.0077	.0402	.9369
7	.0226	.0125	.0276	.3985	.5388
8	.0376	.1028	.7268	.1303	.0025
9	.0526	.0852	.1679	.6792	.0150
10	.0178	.7053	.2694	.0051	.0025

Schwarz (1868) defined the derivative

$$Sf := \frac{f'''}{f'} - \frac{3}{2}\left(\frac{f''}{f'}\right)^2 \qquad [10.1]$$

which is of interest in dynamics because its value indicates the extent to which a process expands to fill a local region of the phase space.

Using successive differencing of a time-series, the total information I in the distribution of the mth differences over k exhaustive mutually exclusive partitions is

$$I^{(m)} := -\sum_{h=1}^{K} p_h log_s(p_h)|_m \qquad [10.2]$$

Using [10.2], suppose that we substitute within [10,1] $I^{(m)}$ values for f^m, so that, for example, $I^{(3)}$ corresponds to f''', employing $I^{(m)}$ summed over the whole range of k subranges in each case. This is to create an entropic analog of the Schwarzian derivative. Then, by definition, ESf (Gregson, 2002) has the form

$$ESf := \frac{I^{(3)}}{I^{(1)}} - \frac{3}{2}\left(\frac{I^{(2)}}{I^{(1)}}\right)^2 \qquad [10.3]$$

Note that [10.3] is defined at least over triplets within a series; the way in which the $I^{(m)}$ is computed drops the actual real measurement units of the data and deliberately loses the first and second moments of the series.

If the data in one subseries are taken as generated by a basic symbolic process X unperturbed by transients, then the probability distribution of the absolute differences of orders one, two, three in that block that are computed for ESf estimation can be used as estimates of $p_i(X|\theta)$, $i = 1, .., 10$ in the Kullback-Leibler expression

$$I(Y, X) = \sum_{i=1}^{5} p_i(Y) \cdot log\left(\frac{p_i(Y)}{X_i|\theta}\right) \qquad [10.4]$$

and that may be taken as a relative mismatch distance, not a true metric distance, because of asymmetry, but a dissimilarity.

Every one of the 10 blocks can in turn be used as the predictor X for the set of 10 Y. The dissimilarity of a set with itself is zero, and the full matrix of the differences are set out in Tables 10.6,10.7, and 10.8. Each row is for a fixed predictor X against which all the 10 Y are compared, running down from X_1 to X_{10}.

These matrices indicate where the variability in information content of the blocks differs without having to commit to a single generating model as theory. Using each block in turn as predictor under some assumptions about error-minimal distributions is an approach reminiscent of Empirical Bayesian statistics. The eigenvalues of these K-L matrices can, in turn, be computed, giving some insight into the role of successive differences in the system's dynamics. If the matrices for differences of different orders 1,2,3 show similar patterns of values, this suggests that the underlying time series are self-similar, which is a property of fractal dynamics.

To compare these three matrices in Tables 10.6, 10.7, 10.8, we may proceed in two ways: just compute the cell-by-cell correlations to see if the matrices treated as 100-term vectors are similar, or find the eigenvectors of each matrix.

Table 10.6: Based on absolute first differences, k = 10.
A Kullback-Leibler matrix of asymmetrical dissimilarities

1	2	3	4	5	6	7	8	9	10
.000	3.242	5.093	.209	.134	5.093	6.567	4.838	6.499	.113
.725	.000	5.217	.557	.650	5.228	4.039	7.774	3.995	.876
1.183	2.631	.000	.915	1.300	.004	5.227	3.215	5.377	1.714
.184	3.719	5.292	.000	.052	5.296	7.024	4.931	7.003	.333
.172	3.603	6.757	.079	.000	6.757	7.421	6.182	7.342	.202
1.185	2.676	.004	.925	1.299	.000	5.395	3.341	5.541	1.728
1.658	.648	4.055	1.309	1.687	4.078	.000	4.188	.055	1.809
1.244	6.632	3.605	.819	1.122	3.617	6.220	.000	6.350	1.506
1.562	.586	5.754	1.280	1.528	5.772	.058	5.861	.000	1.533
.124	2.891	6.308	.365	.180	6.309	5.436	5.852	5.289	.000

The eigenvalues of the first differences matrix Table 10.6 are

$$26.0407, \quad -12.0417, \quad -6.5640 + .3251i, \quad -6.5640 - .3251i \quad -.5446$$
$$-.2028, \quad -.1004, \quad -.0136, \quad -.0083, \quad -.0012$$

in descending order and the last four may be neglected. Note the complex conjugate pair that arises when the matrix is not skew symmetrical about the leading diagonal of zeroes.

**Table 10.7: Based on absolute second differences, k = 10.
A Kullback-Leibler matrix of asymmetrical dissimilarities**

1	2	3	4	5	6	7	8	9	10
.000	2.999	5.111	.442	.241	5.015	6.212	4.548	6.028	.141
.962	.000	5.288	.549	.730	5.321	4.619	8.164	4.594	1.038
1.565	2.809	.000	1.052	1.282	.032	5.580	3.429	5.717	2.000
.345	3.602	5.942	.000	.059	5.911	7.251	5.465	7.209	.421
.205	4.074	5.862	.058	.000	5.790	7.839	5.436	7.754	.251
1.433	3.349	.032	.999	1.150	.000	6.660	3.858	6.751	1.942
1.992	.720	4.871	1.394	1.758	5.029	.000	5.117	.056	1.913
1.287	7.523	4.372	.883	1.061	4.389	7.208	.000	7.281	1.668
1.775	.688	6.293	1.315	1.578	6.399	.056	6.464	.000	1.549
.146	2.612	6.382	.388	.221	6.300	5.720	6.193	5.513	.000

The eigenvalues of the second differences matrix Table 10.7 are

$$28.6073, \quad -13.5271, \quad -6.9312 + .3821i, \quad -6.9312 - .3821i, \quad -.6854$$
$$-.4377, \quad -.0405 + .0094i, \quad -.0405 - .0094i, \quad -.0418, \quad -.0018$$

and now there are two conjugate pairs, but the major terms are almost the same.

The eigenvalues of the third differences matrix Table 10.8 are

$$28.1470, \quad -13.9872, \quad -6.8754, \quad -3.2629 \pm 2.1149i,$$
$$-.6139, \quad -.0434 \pm .0002i, \quad -.0370, \quad -.0024$$

and now the two conjugate pairs are less important, but the total pattern is only slightly changed.

The linear product-moment correlations between Tables 10.6, 10.7 and 10.8 are uninformative, they are $r(6,7) = .9886$, $r(7,8) = .9851$, and $r(6,8) = .9695$. All they can suggest is that the process is self-similar (but not exactly so) at three degrees of differencing, taking Table 10.6 as representing the fastest dynamics. The information in the higher-order difference distributions is generally a little less than that in the first order, the

Table 10.8: Based on absolute third differences, k = 10.
A Kullback-Leibler matrix of asymmetrical dissimilarities

1	2	3	4	5	6	7	8	9	10
.000	2.844	4.562	.295	.359	4.365	5.596	4.178	5.215	.078
.951	.000	5.732	.551	.611	5.823	5.514	8.297	5.506	.963
1.527	3.067	.000	1.155	1.371	.088	6.123	3.650	6.196	1.868
.245	3.250	5.891	.000	.042	5.826	7.011	5.557	6.894	.305
.293	3.078	6.879	.043	.000	6.809	8.046	6.578	7.941	.337
1.318	3.941	.082	1.065	1.254	.000	7.861	4.384	7.819	1.722
1.885	.872	5.629	1.443	1.654	5.923	.000	5.815	.079	1.854
1.244	7.616	4.052	.963	1.165	4.104	7.352	.000	7.350	1.593
1.475	.860	6.169	1.241	1.441	6.328	.087	6.269	.000	1.405
.068	2.290	5.464	.313	.364	5.286	5.361	5.199	4.980	.000

distributions are more skewed towards the lower end. As noted above, the scalar values of the differences, that is their first and second moments, are removed in this analysis, the entropies are scale-independent.

Discussion

As the tremor series can be partitioned approximately into fast and slow dynamics, it is proper to ask where the non-stationarity arises? Taking the evidence of all the relevant tables together, it is seen that the faster dynamics are relatively stable in their evolution, but the slower dynamics fluctuate from block to block. If one thinks of the slower dynamics as a carrier wave, a stable base for supporting faster dynamics sensitive to exogenous effects or to internal nonlinearities, that is *in this case* wrong.

The analysis in Tables 10.6, 10.7 and 10.8 pools over all the 10 blocks, so it is orthogonal to the previous analyses in Tables 10.2, 10.3, 10.4 and 10.5

done block by block. Only the ESf estimates in these four blockwise tables is a hidden bridge between the two orthogonal analyses.

Dynamics of Cardiac Psychophysiology

Some patients undergoing bio-feedback therapy for respiratory disorders do have panic episodes, as they themselves relate. Others do not. The causality and even the detailed dynamics of panic episodes are the subject of considerable controversy (Baker, 1992) and these wider questions are not being entered into here. Rather, our starting point is the fact that palpitations and a local rapid raise in heart rate are the most frequently observed characteristic properties of a panic attack, irrespective of the environmental or intrasubjective situations that appear to trigger the onset. The measures of heart rate activity cited in the relevant literature are usually the rate, in beats/minute, and nothing else. More complex analyses of the sequential dynamics of the heart activity are not usual, partly because the resources available to researchers, both in terms of recording apparatus and mathematical software for subsequent analysis, are relatively recent in their development. The idea that we should investigate the nonlinear dynamics of the heart under conditions when there is no panic, as a possible predictor of panic susceptibility, is virtually untested. At the same time, it is known that cardiac dynamics are intrinsically difficult to model, partly due to their lability and complexity, and contradictory explanations have been advanced for rhythm generation (Dunin-Barkowski, Escobar, Lovering & Orem, 2003).

Gorman and Sloan (2000) review a number of studies in which it is shown that, under non-panic conditions, patients with panic disorders appear to have diminished heart rate variability. In turn, diminished variability is a precursor of myocardial infarction, in short, a good predictor of death, and self-report of panic attacks and higher anxiety can be associated both with lower levels of heart rate variability and increased risk of fatal coronary heart disease (Kawachi, Sparrow & Vokonas, 1994).

This investigation is in two parts: an examination of data from one subject to establish the viability of some of the methodology, and a part in which comparison of some heart beat time series of clinical interest are made using the methods established in the first part.

The first part consists of three epochs, in sequence, each about 5 minutes long. All the three epochs were truncated at 386 interbeat intervals, which time evolution spans approximately five minutes of continuous recording. As the third epoch was deliberately made to be non-stationary, with a brief burst of shallow rapid breathing, the treatment of that series as internally homogeneous is strictly wrong, but is first presented that way, as in the same way as the first two epochs, as an example of the type of error that ignoring nonstationarity or nonlinearity can create (Bunde, Kropp & Schellnhuber, 2002). It is now expected that nonstationarity, better described as the consequences of chaotic itinerancy (Kaneko & Tsuda, 2003), is characteristic of heart activity, where normal dynamic regulation involves return to some attractor basin or basins after transient destabilization (Gorman & Sloan,2000). As has been known for some time, analyses that use linear statistical modelling, either static or with time series assumptions, can obscure the very properties of the dynamics that are both clinically interesting and indeed necessary for the individual's long-term survival (Barndorff-Nielsen, Jensen and Kendall, 1993; Medvinsky, Rusakov, Moskalenko, Fedorov, & Panfilov, 2003).

We note that many of the methods for examining nonlinear time series and their associated dynamics were anticipated in mathematical psychology as long ago as 1960, by Licklider. The main additions now to what was available then centre on chaotic dynamics, the underlying entropy and time series concepts have their origins in the late 19th century. The conventional distributional statistics are as follows, intercept and slope refer to a linear regression over time, the variable in most cases is pulse rate in beats/min. The time series are made up of about 900 interbeat intervals, and normally each of these is about 700-800 millisecs; each starts with the depolarisation peak of the cycle. The nonlinear statistics in Table 10.9 are again LLE, ESf, and ApEn. Low pass and high pass refer to the series after filtering:

Table 10.9: descriptive statistics

measures	Epoch001	Epoch002	Epoch003
min	687	640	300
max	843	847	1011
mean	761.13	770.55	758.60
s.d.	33.69	33.82	117.92
Skewness	.003	-.983	-2.143
Kurtosis	-.643	1.632	5.151
intercept	772	751	708
slope	-.057	+.101	+.221
LLE	.091	.103	.090
Low pass LLE	.147	.130	.089
High pass LLE	.092	.124	.048
ESf	-.7714	-.6722	-.3172
ApEn	1.1035	.9232	.4689

As visual inspection and consideration of what is happening psychophysi-ologically indicate strong serial autocorrelation of the series (except where perturbed), the autocorrelation spectra for each epoch are given in Table 10.10, truncated to 10 lags. It is important here to note that ACF analysis can misidentify a nonlinear process as being i.i.d. (white noise) when in fact it has serial dependencies and may also be nonstationary (Granger, Maasoumi and Racine, 2004).

The serial relations are so simple that a linear lag one model could be fitted if Gaussian residuals were assumed. It is this simplicity that leads to the next suggestion, that a 5-state Markov with weak off-diagonal transi-tion probabilities is predicted (compare Hoppensteadt, 2003). This is ex-plored in Tables 10.11, 10.12, 10.13. The states are each one-fifth of the range of the i.b.i.s, so in, for example, Table 10.13, $(843 - 687)/5 = 31.2$, and the ranges vary for each epoch.

Table 10.10: Serial autocorrelation coefficients

Note: All the coefficients are positive.

lag r	Epoch001	Epoch002	Epoch003
[1]	.9337	.9654	.6692
[2]	.8264	.9152	.5925
[3]	.6858	.8462	.4377
[4]	.5259	.7759	.4627
[5]	.3743	.7087	.3626
[6]	.2383	.6466	.2861
[7]	.1306	.5893	.1915
[8]	.0542	.5362	.2005
[9]	.0113	.4843	.1461
[10]	.0030	.4378	.1543

Table 10.11: Epoch001 Transition Probability Matrix

Note: The last column contains the stationary state vector.

$$EM1 = \begin{bmatrix} \mathbf{.808} & .192 & .000 & .000 & .000 & .123 \\ .091 & \mathbf{.727} & .182 & .000 & .000 & .260 \\ .000 & .135 & \mathbf{.714} & .151 & .000 & .331 \\ .000 & .011 & .191 & \mathbf{.708} & .090 & .234 \\ .000 & .000 & .053 & .368 & \mathbf{.579} & .050 \end{bmatrix}$$

Table 10.12: Epoch002 Transition Probability Matrix

Note: The last column contains the stationary state vector.

$$EM2 = \begin{bmatrix} \mathbf{.840} & .160 & .000 & .000 & .000 & .077 \\ .053 & \mathbf{.787} & .160 & .000 & .000 & .217 \\ .000 & .099 & \mathbf{.763} & .137 & .000 & .338 \\ .000 & .000 & .145 & \mathbf{.794} & .061 & .313 \\ .000 & .000 & .000 & .348 & \mathbf{.652} & .055 \end{bmatrix}$$

Table 10.13: Epoch003 Transition Probability Matrix

Note: The last column contains the stationary state vector.

$$
EM3 = \begin{bmatrix} .550 & .200 & .050 & .150 & .050 & .053 \\ .500 & .000 & .200 & .300 & .000 & .053 \\ .018 & .036 & .727 & .182 & .036 & .144 \\ .004 & .007 & .036 & .931 & .022 & .713 \\ .080 & .080 & .080 & .120 & .640 & .065 \end{bmatrix}
$$

Note, importantly, that there are no absorbing states in the Markovian representation, though some states could be weak attractors, which enables the system to return to some dynamic stability after perturbation, as in Epoch003.

Goldberger (1990, p. 407) remarks that, "A common misconception among dynamicists and clinicians is that the normal heart beats with clockwise regularity. When beat-to-beat heart rate is carefully measured, it is apparent that the healthy heartbeat is quite erratic, even under resting conditions. This variability, however, is subjectively imperceptible and is difficult to assess by routine clinical examination." Goldberger continues, "we have proposed that normal heart-rate variability is regulated by a nonlinear feedback network that generates fluctuations across multiple orders of temporal magnitude, ranging from hours or longer to seconds or less. Furthermore these fluctuations are self-similar. That is, the chaotic-appearing variations apparent on longer time scales are similar to the fluctuations on shorter time scales, although the amplitude of the higher frequencies is lower Preliminary data from our laboratory indicate that ageing is associated with a reduction in fractal dimensionality."

Here we have two sorts of evidence that are compatible with what Goldberger has remarked: the LLE values in Table 10.9 are all positive, indicating that edge-of-chaos activity is very probably present in the dynamics, and the absence of any absorbing states in the Markov matrices. The ESf and ApEn values do covary, this is perhaps the first empirical evidence of this arising as the series are long enough to support the computation of both indices.

Higher Order Statistics, bESf

The series are long enough to permit some exploratory analyses using the higher-order bispectral bESf matrices, that were introduced by Gregson and Leahan (2003). To emphasise, the bold-face entries are within the random surrogate 95% c.i. ranges. Each triangular matrix of the $b(m, n)$ series of triples, $m < n$, is written as

$$
\begin{array}{lllll}
b(1,2) & b(1,3) & b(1,4) & b(1,5) & b(1,6) \\
b(2,3) & b(2,4) & b(2,5) & b(2,6) \\
b(3,4) & b(3,5) & b(3,6) \\
b(4,5) & b(4,6) \\
b(5,6)
\end{array}
$$

Table 10.14: bESf values for Epoch001

$$
EPb1 = \begin{bmatrix}
\mathbf{-.4017} & -.5075 & -.3283 & -.5455 & -.4801 \\
-.5263 & -.4999 & -.4311 & \mathbf{-.4192} \\
-.3482 & -.3372 & -.4440 \\
-.5771 & -.3287 \\
-.4520
\end{bmatrix}
$$

There is little sign of random variations here, in what is essentially a higher-order high-pass filtering, on the higher-frequency fluctuations of the inter-beat-interval. It is expected that Epoch002 will be much the same, but Epoch003 disturbed. Computing in the same fashion:

Table 10.15: bESf values for Epoch002

$$
EPb2 = \begin{bmatrix}
-.3010 & -.4986 & \mathbf{-.4277} & -.5426 & \mathbf{-.4179} \\
-.3064 & -.3381 & -.4643 & -.3507 \\
-.5201 & -.4423 & \mathbf{-.4297} \\
-.4456 & \mathbf{-.4056} \\
\mathbf{-.4205}
\end{bmatrix}
$$

This EPb2 series has more noise than EPb1 in it.

Table 10.16: bESf values for Epoch003

$$EPb3 = \begin{bmatrix} -.3042 & -.4512 & -.4088 & -.3165 & -.5310 \\ -.4105 & -.3483 & -.4247 & \mathbf{-.3785} & \\ \mathbf{-.4338} & -.2446 & -.4811 & & \\ \mathbf{-.4056} & -.3587 & & & \\ -.3967 & & & & \end{bmatrix}$$

The noise has now moved from its location in the previous two matrices. As remarked in our previous work on series of this nature (Gregson, 2003), the bESf matrices are the off-diagonal parts of square skew-symmetric matrices, so their eigenvalues are computable and all real. They provide another way of examining shifts in the dynamics with stimulus disruption; here, dominant stimuli to the cardiac system are breathing rhythms.

Table 10.17: Eigenvalues for bESf for all three Epochs

Eigenvalue	Epoch001	Epoch002	Epoch003
[1]	-2.6538	-2.5340	-2.3636
[2]	-0.2590	-0.1971	-0.2269
[3]	0.2442	0.1735	0.2032
[4]	0.1481	0.1059	0.1366
[5]	-0.1450	-0.0431	-0.0621
[6]	-0.0147	-0.0264	-0.0451

It is clear that the dynamics of all three epochs are not identical and the small differences between the first and third epochs, on the one hand, and the second, are expressed in the lesser [2,3,4] eigenvalues in Table 10.17.

Subdividing the Third Epoch.

Before computing, it is illuminating to examine the graphs of each of the three series, but after they have been high-pass filtered. Figures 10.1, 10.2,

Figures 10.4, 10.5, 10.6: The three cardiac i.b.i. series
Raw data after high-pass filtering

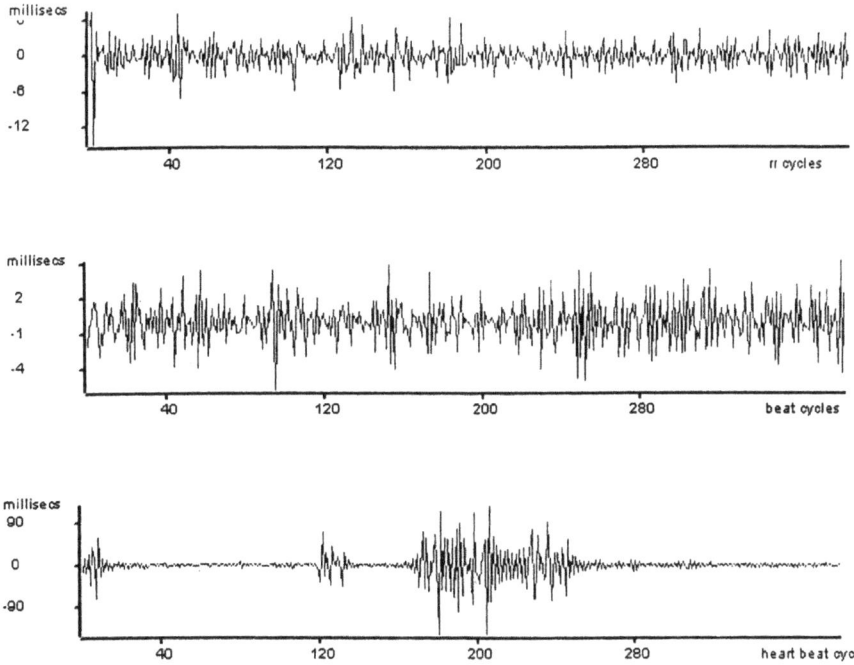

10.3 show the results, the vertical axis scales of the graphs differ to accommodate the different ranges of variability.

We now break the third epoch into the first 160 inter-beat-intervals, then the next 120, then the remaining 106, and compute for each separately; that is, three subseries in sequence. The anomalous subseries, from the high-pass filtered graphs, is the middle one (Epoch032), and the first and third can be considered as controls and should resemble Epochs 001 and 002.

Computationally we may run into difficulties with the shorter series, that may be insufficient to produce stable estimates of LLE or ApEn. For the ESf estimates in Table 10 $\delta y = .01$. These should all be compared with

Epoch003 in Table 10.9.

Table 10.18: descriptive statistics of Epoch003 partitioned

measures	Epoch031	Epoch032	Epoch033
min	300	304	718
max	910	1011	917
mean	738.56	715.63	810.31
s.d.	99.48	161.23	39.88
Skewness	-2.904	1.118	0.162
Kurtosis	8.931	0.587	0.182
intercept	771	576	846
slope	-.417	2.334	-0.698
LLE	.095	.066	.116
Low pass LLE	.039	.038	.118
High pass LLE	.061	.011	.031
ESf	-.4567	-.2566	-.3114
ApEn	.4765	.4064	.4689

The variations in ESf are greater than in ApEn, so it is in this context a more sensitive measure of the destabilization induced in the middle subseries Epoch032 by a temporary burst of a shallow fast breathing rhythm.

Self-Similarity at Different Scales

If the time series have fractal properties, and it is suggested by various authors that they would have, then, if we create a coarser series from the given data by halving the frequency of data points, we should get substantially the same statistical parameters from the analyses in the previous tables. That method is, in fact, called multiscale entropy analysis (Costa, Goldberger and Peng, 2002). As Epoch001 seems to be the most stable, it is this series only that we should employ to check this aspect of the dynamics.

Defining

$$\forall m; \qquad x_m(j) = (x(i) + x(i-1))/2, \qquad j = mod2(i)$$

so that the half frequency series length $n/2$ is 193 double i.b.i.s long the quarter frequency series is 96 long, and the eighth frequency series is 48 long, then ESf(1/2) = -.3111, ESf(1/4) = -.3194, and ESf(1/8) = -.1621.

Multiple analyses on sample series

Comparing the behaviour of different indices computed on successive subsamples of time series is comparing the actions of different filters. A filter may or may not preserve metric information, linear statistics such as autocorrelations or moments do, and the various entropy-based indices do not, but are based in information theory. A filter may reject some information and may also transform it or create spurious information if its inbuilt assumptions are not compatible with the data structure. If the basic processes being studied are not stationary, then a filter that only represents some of the information present may vary in what it captures.

As an illustration of the variability that can be found under conditions that might reasonably be thought to be stable and even tranquil, Table 10.19 shows one series collected under meditation conditions. The values are here expressed as interbeat intervals (i.b.i) in 100ths of a second. This series could be taken as one sort of base rate situation. The various measures are not usually all computed on the same data set and show some mutual disagreements. All entropy-based measures I have seen have inbuilt tuning parameters to be set in their computation and their settings can critically affect the performance of the measures, that is filters, as process identifiers (Richman and Moorman, 2000).

The relative utility of the various measures is not simply decided, as they reflect different properties of the nonlinear dynamics. We can say, using the LLE as the basic criterion, that all the series are chaotic, to a small degree, and that the largest Lyapunov does vary a bit over time in what may be psychophysiologically interpreted as a stable process, as has been reported in our previous work (Gregson and Leahan, 2003). ESf does have a stronger relation to LLE than ApEn, which is not surprising because ESf uses more series information to construct its values.

Figures 10.7, 10.8, 10.9: The raw data of Meditation blocks 1, 4 and 10
Showing non-stationarity. Vertical scales vary.

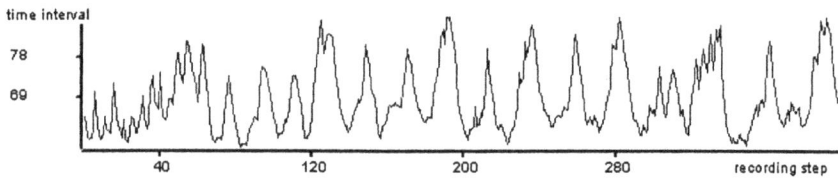

Meditation block 1

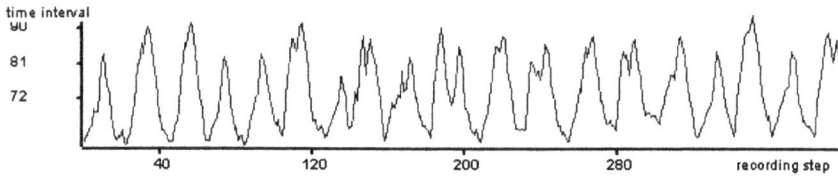

Meditation block 4

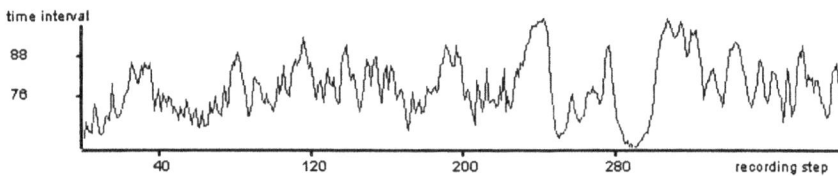

Meditation block 10

Each of the ESf values has been checked against the 95% confidence interval found from a set of 50 random surrogate series, and shown to be outside and above the associated random interval.

The higher-order bESf matrices for two selected blocks are shown in Tables 10.20 and 10.21. As previously, the bold-face entries are within the random surrogate 95% c.i. The mixture of chaotic and random components of the higher-order dynamics again appears. There is a possible resemblance between Tables 10.8 and 10.13, interestingly as one is drawn from Australia and the other from the USA, they are independent analyses.

Table 10.19: Descriptive statistics of Meditation series
This series of ECG data is taken from the PhysioNet archives.

Block	LLE	mean	Kurt	ESf	D2	H	ApEn
1-400	+.1123	68.88	-0.572	-.4341	1.813	.1866	1.020
401-800	+.1062	71.53	-0.999	-.4678	1.853	.1472	0.950
801-1200	+.1198	73.51	-1.287	-.4594	1.819	.1805	0.829
1201-1600	+.0986	73.48	-1.135	-.4529	1.801	.1987	0.953
1601-2000	+.1285	73.70	-0.973	-.4837	1.767	.2333	0.949
2001-2400	+.1513	74.63	-1.092	-.4935	1.832	.1684	1.029
2401-2800	+.1224	74.12	-1.047	-.4238	1.807	.1931	0.954
2801-3200	+.1496	75.02	-1.188	-.4359	1.810	.1934	0.872
3201-3600	+.1284	73.38	-0.622	-.4857	1.859	.1411	1.053
3601-4000	+.1254	75.85	-0.403	-.4185	1.769	.2314	1.111
–	–	–	–	–	–	–	–
cftv	.129	.025	.300	.056	.016	.155	.082
regr	1	63.53	-2.02	-0.34	-0.02	0.06	0.15
corr	1	.544	-.116	-.214	-.010	.034	.030

Key: Block: beats in sequence. LLE: largest Lyapunov exponent. mean: average i.b.i. Kurt: kurtosis. ESf: entropic analogue of Schwarzian derivative. D2: fractal dimensionality. H: Hurst index. ApEn: approximate entropy ($k = 10$, $\delta = .02$). cftv: coefficient of variation ($\sigma/|\mu|$). regr: slope of regression of variable on LLE. corr: product-moment correlation of LLE with variable.

Whereas the series of i.b.i. durations under meditation was a time interval series, or a point process, the next example is a fixed time interval variable value series, of mv of cardiac activity at 100 Hz under conditions of apnea. The intervals between peak values in the graphs, which are not exactly constant, are the same in meaning as the variable i.b.i.s in the meditation example. That is, the scale here is enhanced and Table 10.19 is based on slow dynamics while Table 10.22 reflects the higher frequency components within each beat cycle, at 100 times the previous scale. It does not follow that, because the i.b.i. series in Table 10.19 shows signs of chaotic dynamics, the same would necessarily be expected of the mv series in Table 10.22. The analysis follows exactly the same steps. It is seen that, though the LLE values are still all positive, they are much lower than in Table 10.19 and, given the relatively short series, may be treated as virtually zero. In apnea, the heart action is compromised, which is a dampening of the dynamics. It is not possible directly to compare ESf or ApEn values in Tables 10.19 and 10.22, because the two time series are quite different in their meaning, but one may note the variability of ESF within each series. In Table 10.22 the coefficient of variation increases from .056 to .124, whereas for ApEn it decreases from .082 to .056.

Table 10.20: bESf values for Block 4: with LLE lowest

$$
EPbMd4 = \begin{bmatrix}
\mathbf{-.4373} & \mathbf{-.4315} & -.2604 & -.3918 & -.3996 \\
-.6105 & -.5100 & -.2872 & \mathbf{-.3420} & \\
-.2901 & -.4853 & -.1745 & & \\
\mathbf{-.3526} & -.2726 & & & \\
-.2026 & & & &
\end{bmatrix}
$$

Table 10.21: bESf values for Block 6: with LLE highest

$$
EPbMd6 = \begin{bmatrix}
-.7213 & -.4520 & -.5510 & -.3137 & \mathbf{-.4894} \\
-.3603 & -.6031 & -.4299 & -.2864 & \\
\mathbf{-.4881} & -.4290 & -.5306 & & \\
\mathbf{-.4580} & -.4484 & & & \\
-.4372 & & & &
\end{bmatrix}
$$

The original data did not specify if medication was being used when these records were obtained, so it is not possible to partial out any causality, digitalis is often used to control arrhythmias, and these series are quite regular and have a recurrent defined periodic spectrum, with no evidence of tachycardia.

Here, the ApEn values are very stable, but the ESf shows fluctuations, it uses higher-order derivatives and these appear to play a role that is not detected by ApEn. The difference in the LLE values between Tables 10.19 and 10.22, suggesting that the slow i.b.i. series are chaotic, but the high frequency mv series within the i.b.i.s are not chaotic, is not novel, but has been considered for brain dynamics at different levels (Wright and Liley, 1996; Gregson, 1997).

Tables 10.23 and 10.24, unlike Tables 10.20 and 10.21, are generated by the high frequency dynamics that produce the p,q,r,s fluctuations in mv in each cycle near the peak. The values within the triangular bESf matrices are lower for this (fast) series, than for the (slow) meditation i.b.i.s. There is little evidence of any random components here, so the process is nonlinear but not chaotic at a lower (slow) level, taking all the evidence in Table 10.24 together.

Panic Pre-dynamics

The findings of Gorman & Sloan (2000) on the importance of heart rate variability have already been mentioned and it is important to note that such variability itself is complex and has sources that are usually confounded without appropriate statistical analyses (Chon, Zhong, Wang, Ju and Jan, 2006.

But the question remains, what measures of variability in time series are most relevant? Tucker, Adamson & Mirander (1997) use power spectral analysis. That presumes linearity and dynamic stationarity, and thus filters out aspects of the dynamics that are known to be present in normal cardiac activity, that is, in fact, edge-of-chaos (Winfree, 1987). So far as can

Figures 10.10, 10.11, 10.12: The raw data of Apnea mv blocks 1, 5 and 7
Illustrating cyclic variation. Vertical scales vary

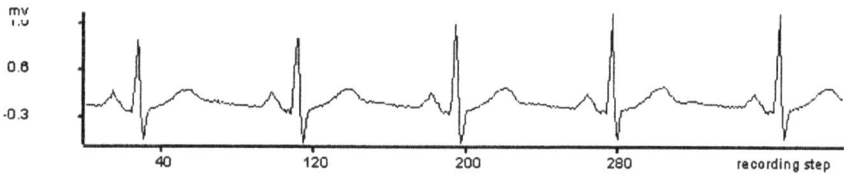

Apnea block 1

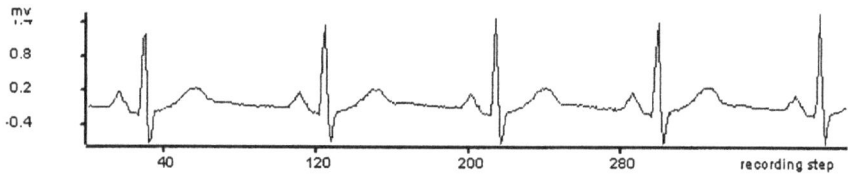

Apnea block 5

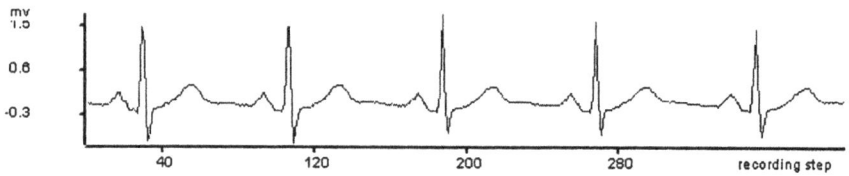

Apnea block 7

Table 10.22: Descriptive statistics of Apnea mv series
This series of ECG data is taken from the PhysioNet archives.

Block	LLE	mean	Kurt	ESf	D2	H	ApEn
1-400	+.0246	.0009	13.81	-.4771	1.863	.1373	.383
401-800	+.0328	.0043	18.90	-.3427	1.741	.2594	.326
801-1200	+.0481	-.0021	19.25	-.4562	1.854	.1458	.344
1201-1600	+.0163	.0013	19.02	-.3357	1.900	.1003	.323
1601-2000	+.0625	-.0043	13.45	-.4150	1.860	.1404	.363
2001-2400	+.0516	.0032	17.69	-.4270	1.841	.1594	.371
2401-2800	+.0645	.0002	13.49	-.4413	1.876	.1244	.370
2801-3200	+.0387	-.0047	14.85	-.3303	1.863	.1370	.359
3201-3600	+.0439	.0041	15.24	-.4348	1.824	.1762	.379
3601-4000	+.0262	-.0077	16.19	-.4430	1.835	.1648	.370
–	–	–	–	–	–	–	–
cftv	.375	8.070	.138	.124	.022	.262	.056
regr	1	-.006	-58.355	-1.021	.135	-.131	.436
corr	1	.0245	.4001	-.3071	.0513	.0498	.3349

Key: As Table 19. For ESf $k = 10, \delta = .0025$

Table 10.23 bESf values for Block (3) with ESf high

$$EPbApn3 = \begin{bmatrix} -.3243 & -.3653 & -.1395 & -.2625 & -.1647 \\ \mathbf{-.3043} & -.3076 & -.2130 & -.4101 & \\ -.4350 & -.2807 & -.3520 & & \\ -.1827 & -.3461 & & & \\ -.1582 & & & & \end{bmatrix}$$

Table 10.24: bESf values for Block (8): with ESf lowest

$$EPbApn8 = \begin{bmatrix} -.3027 & -.4218 & -.2350 & -.5100 & -.3702 \\ -.3523 & -.2472 & -.5195 & -.3288 \\ -.4834 & \textbf{-.4099} & -.4909 \\ -.2613 & -.5160 \\ -.1878 \end{bmatrix}$$

be deduced from literature reviews, variability is in most studies simply equated with Gaussian variance about a mean heart rate.

Though the focus of this intended study is on heart rate variability during periods when patients are not having panic, there are still other conceptual difficulties to be noted. Margraf and Ehlers (1992, p. 154) remark that it is inadequate to "dichotomize the subjects' complex responses simply into panic or not panic". They note that it is not the case that during a panic attack a subject's heart rate always increases. "In addition to being prone to biases, relying on the artificial dichotomous variable 'panic/not panic' entails a considerable loss of information compared to the measurement of continuous variables." The same argument can be applied to the time series analysis of cardiac activity, if the latter is only encoded as "elevated/normal" in the local mean rate.

Margraf and Ehlers conclude (1992, p. 224) that, "a pure medical illness approach to panic has been shown insufficient, possible biological vulnerabilities have been revealed only for subgroups, and the necessity of a psychophysiological perspective has been underlined the integration of different levels of analysis is a basic problem of all modern neuroscience. Rather than biological or cognitive reductionism a true psychobiological perspective is needed."

Notes

The statistical calculations were made partly by using SANTIS software from the Physiology Department of the University of Aachen and by the author's own programs in Linux Fort77, and in Mathematica. Dr Pincus kindly provided code of his Approximate Entropy program which was rewritten into Linux. The time series in the second part were drawn from the PhysioNet Archives.

Appendix

For $\mu = .735$ the eigenvalues of LS are, in complex form:

$$1.0, \ .715 + .1009i, \ .715 - .1009i, \ .645 + .1645i, \ .645 - .1645i$$

The dimensionality of the embedding space of the Shilnikov attractor is three. A perspective picture of its trajectory is given by Mitchener & Nowak (2004, p. 704). The formal mathematics of the Shilnikov phenomenon are given by Wiggins (1990, pp. 573, 602). It arises in the dynamics near an orbit homoclinic to a fixed point of saddle-focus type in a third-order ordinary differential equation. The equation can be of the form

$$
\begin{aligned}
\dot{x} &= \rho x - \omega y + P(x, y, z), \\
\dot{y} &= \omega x + \rho y + Q(x, y, z), \\
\dot{z} &= \lambda z + R(x, y, z).
\end{aligned}
$$

where P, Q, R are complex. The eigenvalues are given by $\rho \pm i\omega$.

A deeper mathematical treatment is given by Wiggins (1988, p. 227 et seq), wherein it is noted that the attractor possesses a 2-dimensional stable manifold and a 1-dimensional unstable manifold. The time of flight between the two attractors is a function of λ.

The dynamics in the local regions of saddle-nodes, of which this can be an example, are treated qualitatively by Shilnikov, Shilnikov, Turaev & Chua (1998). A further treatment of the role of eigenvalues in the subsystem identification of metastable Markov systems has recently been reviewed by Huisinga, Meyn and Schütte (2004).

Chapter 11

Appendix: Nonlinear Psychophysical Dynamics

This appendix summarises some algebraic properties used in nonlinear psychophysics with reference to their contextual literature in pure mathematics. It enlarges on some material presented graphically in Chapter 1, and drawn on particularly in Chapters 3, 7, 8 and 10.

Alternative representations of the dynamics in real time:

(1) Julia sets of the dynamics, with coordinates the starting points $Y(Re, Im)_0$ of the recursive iterations which yield the trajectories; attractor basins and self-similarities displayed with magnification.

(2) Dynamic partitions of the system's parameter space; local regions associated with attractor characteristics. Coordinates are the equation's fixed parameters, not the internal variable.

(3) Mappings indicating homoclinic and heteroclinic orbits, distinguishing real and complex spaces, and Poincaré space sections.

(4) Trajectories as parallel time series of Real and Imaginary parts of the system variable Y, recurrence maps, and alternative series based on transformations into polar coordinates.

(5) Symbolic dynamic encodings of the time series created by discrete partitioning into Markovian state variables. These are the basis used here to identify arpeggio-like subsequences.

(6) Response surfaces of terminal values of the trajectories over the parameter space. These come in pairs and one in the Reals corresponds most closely to psychophysical data from the perspective of an external observer. Surfaces may be differentiated and the derivatives used to estimate response latency frequency distributions via Jacobians.

(7) Cross-entropy maps from input parameter values to Real parts of the output at a fixed number of iterations. Used to assess information loss in stimulus-response pairings.

(8) Time series of convoluted trajectories, with variable gain parameters, matching in meaning $< 4, 5, 6, 7 >$.

A general mathematical results which Milnor (1992) draws on is that any cubic polynomial map from the complex numbers C to C is conjugate, under a complex affine change of variable, to a map of the form

$$z \mapsto f(z) = z^3 - 3a^2 z + b \qquad [11.1]$$

and we note that Γ in nonlinear psychophysical dynamics (NPD)(Gregson, 1988, 1992, 1995) which is defined as

$$\Gamma: \quad Y_{j+1} = -a(Y_j - 1)(Y_j - ie)(Y_j + ie), \quad i = \sqrt{-1}, \quad j = 1, ..., \eta \qquad [11.2]$$

falls into this family.

A more general form which exemplifies the structural questions involved is to rewrite Γ as its generic family

$$\Gamma: \quad Y_{j+1} = f(\pi, ..., \pi_\zeta, Y_{j-k}(Re, Im), Y_j(Re, Im))$$

in which π is any parameter, and the presence of an indeterminate number of complex eigenvalues has to be allowed a priori. The dynamics vary critically with the eigenvalue structure (Ilyashenko & Li, 1999). In [11.2] we have simply $\zeta = 2$ and one complex eigenvalue, in other words, a complex conjugate pair. The subscript k allows for delay, and in [11.2] is zero, but has been 1 in some cases considered (Gregson, 1995, 1998). The terminating parameter η in [11.2] is not a necessary part of the definition; it is merely an arbitrary limit to the trajectory.

The parameters $\{\pi\}$ are usually taken to be scalar constants, but in Γ_c convolutions become variables under some definable conditions. The shape of the plot of $Y(Re)/a$ is similar to a cumulative normal ogive, and also to

$$v(x) = \begin{cases} 0, & x \leq 0, \\ sin^2 \frac{\pi}{2}x, & 0 \leq x \leq 1, \\ 1, & x \geq 1. \end{cases}$$

(Daubechies, 1992, p.74).

If, instead, the situation were to be modelled in stochastic dynamics, then the nearest statistical model would be the *random walk plus noise hierarchical form* (Atkinson, 1998)

$$y_t = \alpha_t + \epsilon_t \quad \epsilon_t \sim N(0, \sigma_\epsilon^2) \tag{11.3}$$

$$\epsilon_{t+1} = \alpha_t + \eta_t \quad \eta_t \sim N(0, \sigma_\eta^2) \tag{11.4}$$

where y replaces $Y(\text{Re})$ in the NPD structure, and α replaces a in Γ. Both the nonlinear dynamical convoluted system and the stochastic random walk system use a single time counting unit, as usually presented, both for the driving and the response parts, which is t in [11.3,11.4] above and j in [11.2].

Consider a single channel psychophysical process mediated by a locally destabilised Γ trajectory (Gregson, 1988). There are three successive time segments called epochs,

$$t = 1, ..., n_1, n_1 + 1, ..., n_2, n_2 + 1, ..., N$$

The stimulus is only made to be present in the middle time segment, of duration $n_2 - n_1$. The gain settings are a_g in the first and third epochs, and a_f in the middle epoch. To reflect familiar psychological terms, g is for *ground* and f is for *figure*. For the ground condition in epochs 1,3 Γ_g is [11.2] with $a = a_g$, and for the stimulus epoch Γ_f is [11.2] with $a = a_f$.

Converting to polar coordinates, with starting values Y_0(Re,Im), set

$$x = Y \text{ (Re)}, \quad w = (Y_0(\text{Im}).10^c)^{-1}, \quad y = w(Y \text{ (Im)}.10^{2c}) \qquad [11.5]$$

then

$$r = (x^2 + y^2)^{1/2}, \qquad tan\theta = y/x \qquad [11.6]$$

are the renormed polar coordinates required.

For example, in studying the dynamics of rhythmic motion, Treffner and Turvey (1996) use

$$d\phi/dt = \Delta\omega - a \sin(\phi) - 2b \sin(2\phi) + \sqrt{Q}\zeta_t \qquad [11.7]$$

to predict equilibria and fluctuations in symmetry breaking during hand coordination tasks.

In general, the family of maps

$$x \mapsto x + \mu + \epsilon \cos 2\pi x = f(x, \mu, \epsilon), \qquad x \in R^1, \epsilon \geq 0 \qquad [11.8]$$

where points in R^1 that differ by an integer so that [11.8] can be regarded as a map defined on the circle $S^1 = R^1/Z$ is a generator of Arnol'd tongues due to the presence of terms in $O(\epsilon^3)$. So far, there exists no formal proof that [11.2] with periodicity in a resembles a case of [11.8], but see Chavoya-Aceves et al (1985)[1]

Milnor (1992, p.11) illustrates this point by comparing what he calls "pointed-swallow" configurations in Arnol'd tongues with arch configurations in cubic maps.

---

[1] Arnol'd tongues can be generated also by the sine-map of the circle, which is

$$Y_{n+1} = Y_n + a + b \sin 2\pi Y_n (mod 1),$$

effectively the same form as [11.8].

Let $a = f(\alpha| \sin(\kappa\pi t)|), 2.5 < a < 5.0, t \equiv j$ within the stimulus epoch. Here κ determines the periodicity and α determines the amplitude, which has to be constrained to keep the dynamics within the attractor basin of Γ.

Table 11.1: Local behaviour of output after a sinusoidal peak input
r_t for various e after 50 interations of Γ

$e \rightarrow$	.2	.25	.3	.35
$a \downarrow$				
4.3920	2.5663	0.9946	0.7954	4.9295
4.6789	3.9077	2.5827	1.4365	6.5242
4.8790	1.6367	2.4472	5.6596	13.9506
4.9842	0.7577	4.2730	9.4996	23.4724
4.9904	0.9428	4.8962	18.8983	55.2053
4.8973	1.9336	8.3213	23.8296	66.6827
4.7086	2.2313	2.0982	52.4784	203.6165
4.4319	3.4945	0.9261	10.1625	54.1295

Some of the results here have been foreshadowed by Ott, Grebogi and Yorke (1990). They write: "The behaviour of the response system depends on the behaviour of the drive system, but the drive system is not influenced by the response system. We have split the drive system into two parts. The first part represents the variables that are not involved in driving the response system, and the second part represents the variables that are actually involved in driving the response system." In cascaded Γ notation this is exactly the same thing as putting $a_{k+1} = f(Y_{\eta|k}(\text{Re}))$ (Gregson, 1995, gives detailed examples).

Guided by psychophysical considerations, the situation chosen for exploration involves setting up a convolution of one Γ trajectory onto a cascade of a second series of Γ trajectories, where the first Γ_1 is quasiperiodic or chaotic in the reals, and using

$$a_2 = \alpha + \beta Y_1 \text{ (Re)} \qquad [11.9]$$

where α and β are real positive scalars to make the range of $Y_2(\text{Re})$ lie within stable limits,, as a_2 is input to Γ_2 with its η_2 not less than 4. This convolution, $\Gamma_C = \Gamma_2(\Gamma_1)$ in operator notation, with settings of e chosen to give autonomous dynamics beyond the first bifurcation for each Γ, is of potential interest and psychophysically possible. Such convoluted and cascaded series are also compatible in principle with what is known about pathways between the hippocampus and the frontal cortex (Halgren & Chauvel, 1996).

For illustration consider two examples, for complex and for polar trajectories. Suppose that $a_1 = 4.1251, e = .447226, \alpha = 2.13$ and $\beta = 2.55$, then $Y_1(\text{Re})$ runs into quasiperiod 8 with roughly to the first decimal place the sequence $.9\bullet, .3\bullet, .8\bullet, .5\bullet, .9\bullet, .2\bullet, ...,$ (the symbol $\bullet$ in each step indicates omitted digits in the recursion of Γ). The a_2 series then starts with $4.30\bullet, 3.25\bullet, 3.78\bullet, 3.49\bullet, 3.73\bullet, 3.68\bullet, ..$ and the generated $Y_2(\text{Re})$ series with $\eta_2 = 5$ runs onto quasiperiod 8 with $.9\bullet, .7\bullet, .8\bullet, .6\bullet, .9\bullet, .7\bullet, .9\bullet, .6\bullet...$ But in the first phase of the convolution the largest Lyapunov exponent is negative, for the $Y_1(\text{Re})$ series, it moves from positive to negative for the polar r_1 series, it is positive and slowly stabilises for $Y_2(\text{Re})$ and is positive and unstable for polar r_2.

A second case where the dynamics are very near to explosion and have become aperiodic can be created by setting $a_1 = 4.132503, e = .447226, \eta_1 = 360$, then $\alpha = 2.13, \beta = 2.55$, and $\eta_2 = 4$, which creates an interesting series in $Y_2(\text{Re})$. The $Y_1(\text{Re})$ series is quasiperiodic 8, with $.9\bullet, .3\bullet, .8\bullet, .5\bullet, .9\bullet, .2\bullet, .8\bullet, .6\bullet, ..$ and to characterise the $Y_2(\text{Re})$ or r_2 series a variant of symbolic dynamics is useful (Kantz and Schreiber, 1997).

From examination of the generated $Y_2(\text{Re})$ series for $t = 1, ..., 359$ a convenient partitioning of Y to create an X variable is made (the set of values $\{X\}$ become the states of a Markov process, as are used in the box-counting methods for mapping attractors; see Kreuzer, 1987).

and a series is then of the form $X_1 m_1 X_2 m_2 X_3 m_3 .... X_k m_k ..$ where $m_k = 0, ..., \infty$ is the number of successive $\emptyset$ values of X in the inter X interval $X_k X_{k+1}, X \in \{A, B, C, D, E, F\}$. For example a continuous segment of the series is found to read as

.......E3C3F1 **C1C3E1A1**C3F1D1C3D1B1C3F1 **C1C3E1A1**C3F1 $\rightarrow$
C1C3E2C3F1 **C1C3E1A1**C3F1D1C3...   [11.10]
The $\rightarrow$ indicates unbroken continuity. The subsequence set in boldface

---

Table 11.2: Partitioning of input to create a symbolic variable

$$\text{if } 0 < Y < .61 \quad X \equiv \varnothing$$
$$\text{if } .61 < Y < .62 \quad X \equiv A$$
$$\text{if } .62 < Y < .63 \quad X \equiv B$$
$$\text{if } .63 < Y < .64 \quad X \equiv C$$
$$\text{if } .64 < Y < .65 \quad X \equiv D$$
$$\text{if } .65 < Y < .66 \quad X \equiv E$$
$$\text{and if } Y > .66 \quad X \equiv F$$

---

may be regarded as an arpeggio-like event Φ. There are two time series here which may provide evidence of near-chaos or chaotic dynamics; the inter-Φ-interval frequency distribution and its series, its first two terms being 16 and 14 Xm steps, and the series of m values, at a finer resolution in time, which is running as

.....3,1,1,3,1,1,3,1,1,3,1,1,3,1,1,3,1,1,3,1,3,2,3,1,1,3,1,1,3,1,1.. [11.11]

which is almost period 3.

The symbolic dynamics just constructed rest on only the $Y_2(\text{Re})$ series; if instead the associated polar modulus r_2 series is used, not quite the same patterns merge. The comparison analysis is performed on the same segment of the trajectory of Γ_2. Comparing with [11.10], but for polar r_2, reading the iterations from the same time point onwards,

...EC2C2FF1C1A2E1A1C3FFFFC2FFF1FC2FEF1FC2F $\rightarrow$

F1C1CC2E1AFCD2F1C1C2.. [11.12] from which it is not possible to identify any recurrent Φ within the same time window. The new sequence of m values, corresponding to [11],is now

....2,2,0,1,1,0,2,1,1,3,0,0,0,0,2,0,0,0,1,0,2,0,0,1,0,2,0,1,1,1,0,2,0,0,0,2,.. [11.13]

Making a change in e from 0.447226 to 0.47227 is a sufficient example, still just within stability retaining a_2 unchanged.

For $Y(\text{Re})$ over the same range of iterations as before we now have (compare [11.10]):

...E3C3E1A1C3F1C1C3E3C3 **F1B1C3E1A1C3**$\rightarrow$

F1B1C3FB1C3 F1B1C..   [11.14]

and again there is the emergence of an arpeggio Φ but not the same one as in [11.10]. The m series corresponding to [11.11] is like

...3,3,1,1,3,1,1,3,3,3,1,1,3,1,1,3,1,1,3,1,1,3,1,1,...   [11.15]

which is apparently quasiperiodic 3, rather like [11.10]. Repeating the exercise for the corresponding symbolic dynamics of polar modulus r_2 yields (compare [11.12])

......EE1FC3E1A1CF1FF1CFCE2E2FCC1 **FF1**B1C2FEFA1→

C2 **FF1**B1C3F1BFCF1 **FF1B**FCF1..   [11.16]

and the weak emergence of a Φ is seen. The corresponding m series is now

....0,1,0,3,1,1,0,1,0,0,0,0,2,2,0,0,1,0,1,1,2,0,0,0,1,2,0,1,1,0,1,0,0,0,1,0,1...

[11.17]

which appears to be locally aperiodic.

It is immediately obvious that the autocorrelation structure of the response Y_2(Re) series can be different for the two cases, and the autocorrelation structure will be related to the largest Lyapunov exponent. For illustration, recomputing the last example given, in [11.14-11.17], with $a = 4.132503, e = .447227$, which is period 8 approximately during the driving stage, gives (for comparison with [11.14]) for Y_2(Re),

......EFC1 **CF2**DF2A1 **CF2**EFC1 **CF2**DFA **CF2**EFB1 **CF2**DFA→

CF2EFB1 **CF2**DFB **CF2**EFB1CF...   [11.18]

and in this string we can identify two Φ which may later coalesce. The corresponding m series (compare [11.17]) is thus

0,0,1,0,2,0,2,0,2,0,0,1,0,2,0,0,1,0,1,0,0,1,0,2,0,0,1,0,2,0,0,1,0,2,0,0,1,0,2,.

[11.19]

which is period 5 after it stabilizes. The new polar r_2 series is quasiperiodic 8 and quite uninteresting.

The definition of mutual information $I(X,Y)$ takes a general form, with $\forall i \; x_i \in X, \quad \forall j \; y_j \in Y$, and bivariate summation implicit

$$I(X,Y) = \sum_{i=1}^{T} \sum_{j=1}^{T} p_{XY}(x_i y_j) log_2 \left(\frac{p_{XY}(x_i y_j)}{p_X(x_i) p_Y(y_j)} \right) \qquad [11.20]$$

The form of [11.20] extends to cover the case where one series is lagged onto itself, or delayed onto another realization of the same series, by writing (Abarbanel, 1996, p.28), where the lagged delay is θ recursive steps of the map,

$$I(\theta) = \sum_{n=1}^{T} (p(y_n, y_{n+\theta}) log_2 \left(\frac{p(y_n)}{p(y_n)p(y_{n+\theta})} \right) \qquad [11.21]$$

and we may then proceed to find the θ which maximises $I(\theta)$ for a given realization of the process. For example, if [11.14] is repartitioned at values $Y(\text{Re}) = .54,.57,.62,.63,.64,.66$ to give seven states over the continuum 0,1 then it is seen, as the system is stationary in its parameters, that it is near equilibrium with a stationary state probability vector $.362,.139,.140,.008,.137,.002,.108$. It is this vector which is stochastically externally observable behaviour; the actual values depend on where the arbitrary partitions on Y are set and how many are chosen. The basic information definition $H = -\sum_{i=1}^{7} p_i log_2(p_i)$ can be computed for the vector.

A slightly different exploration of the $Y_2(\text{Re})$ series, for various η_2, is used in Table 11.3. It is suspected that the information of the $Y_2(\text{Re})$ frequency distribution is a function of η_2, from some precedents. For Table

---

Table 11.3: Information capacity of the $Y_{2,\eta}(\text{Re})$ distributions
For various iterations, of the original and the first and second differenced series.

η_2	$H[Y(Re)]$	$H\Delta^1[Y(Re)]$	$H\Delta^2[Y(Re)]$
2	2.502	1.419	1.376
4	2.505	1.470	1.418
5	2.508	1.423	1.422
6	2.500	1.367	1.307
7	2.529	1.447	1.457
8	2.500	1.369	1.307
10	2.500	1.351	1.309
11	2.501	1.351	1.309
15	2.501	1.326	1.312

11.3 the Y_2(Re) range has been partitioned into 10 equal steps of 0.1; if the distribution were rectangular then H would be 3.322 ($= log_2 10$). Actually, it is not ever that, and for Y_1(Re) $H = 2.500$, so, except for an apparent irregularity at $\eta_2 = 7$, the information is not much altered by the convolution, though the actual distribution of Y values is altered. The first Lyapunov exponent can also reverse sign, and is not the same for the first and second differences (Δ^1, Δ^2) of the Y_2 series.

A relevant comment of Boon and Prigogine (1998, p.4) who write:

> Le phénomène musical possède une certaine cohérence, intermédiare entre l'aléatoire et le périodique: il y a des corrélations, mais pas de structure strictement déterministe La musique possède une symétrie brisée: elle se déroule selon un ordre temporel

Notes

(1) The standard notation C, R, Z refers respectively to Complex, Real and Integer domains. The notation of NPD algebra is made to be fully consistent with the previous text references given.

(2) The idea of arpeggios Φ in the symbolic dynamics used here apparently parallels what are also called isolating segments in chaotic phase sequences (Wójcik, 1998).

References and Bibliography

Abarbanel, H. D. L. (1996) *Analysis of Observed Chaotic Data.* New York: Springer-Verlag.

Abraham, N. B., Albano, A. M., Passamante, A. and Rapp, P. E. (Eds.) (1989) *Measures of Complexity and Chaos.* New York: Plenum Press.

Abrams, D. M. and Strogatz, S. H. (2006) Chimera States in a Ring of Nonlocally Coupled Oscillators. *International Journal of Bifurcation and Chaos, 16,* 21-37.

Aczél, J. (1966) *Lectures on Functional Equations and their Applications.* New York: Academic Press.

Adami, C.,and Cerf, N. J. (2000) Physical complexity of symbolic sequences. *Physica D, 137,* 62-69.

Aguirre, L. A., and Billings, S. A (1995) Identification of models for chaotic systems from noisy data: implications for performance and nonlinear filtering. *Physica D, 85,* 239-258.

Ambühl, B., Dünki, R. and Ciompi, L. (1992) Dynamical Systems and the Development of Schizophrenic Symptoms - An Approach to a Formalization. *In:* Tschacher, W., Schiepek, G. and Brunner, E. J. (Eds.) *Self-Organization and Clinical Psychology.* Berlin: Springer-Verlag. 195-212.

Amit, D. J. (1989) *Modelling Brain Function: The World of Attractor Neural Networks.* New York: Cambridge University Press.

Andrews, D. F. and Herzberg, A. M. (1985) *Data.* [Springer Series in Statistics.] New York: Springer-Verlag.

Arecchi, F. T. (1987) Hyperchaos and 1/f spectra in nonlinear dynamics: the Buridanus donkey. *In:* Caianiello, E. R. (Ed.) *Physics of Cognitive Processes.* Singapore: World Scientific. 35-50.

Arecchi, F. T., Badii, R. and Politi, A. (1984) Low-frequency phenomena in dynamical systems with many attractors. *Physical Review A, 29,* 1006-1009.

Argoul, F. and Arneodo, A. (1986) Lyapunov Exponents and Phase Transitions in Dynamical Systems. *In:* Arnold. L. and Wihstutz, V. (Eds.) *Lyapunov Exponents. Lecture Notes in Mathematics, No. 1186* Berlin: Springer-Verlag.

Århem, P., Blomberg, C., and Liljenstrom, H (1999) *Disorder versus Order in Brain Function.* Singapore: World Scientific.

Atkinson, A. C. (1998) Discussion of the paper by Hodges. *Journal of the Royal Statistical Society, Series B, 60,* 521.

Attneave, F. (1959) *Applications of Information Theory to Psychology.* New York: Holt, Rinehart and Winston.

Baddeley, A. D. and Logie, B.A. (1999) Working memory: The multiple component model. *In:* A. Miyake and P. Shah (Eds.), *Models of working memory: Mechanisms of active maintenance and executive control.* Cambridge, UK.: Cambridge University Press. 28-61.

Badii, R. and Broggi, G. (1989) Hierarchies of relations between partial dimensions and local expansion rates in strange attractors. *In:* Abraham, N. B., Albano, A. M., Passamante, A. and Rapp, P. E. (1989) *Measures of Complexity and Chaos.* New York: Plenum Press. 63-73.

Baier, G. and Klein, M. (Eds.) (1991) *A Chaotic Hierarchy.* Singapore: World Scientific.

Baker, R. (Ed.) (1992) *Panic Disorder: Theory Research and Therapy.* New York: John Wiley and Sons.

Ball, R. (Ed.) (2003) *Nonlinear Dynamics from Lasers to Butterflies. World Scientific Lecture Notes in Complex Systems, Vol. 1* Singapore: World Scientific.

Bar-Hillel, M. and Wagenaar, W. A. (1991) The Perception of Randomness. *Advances in Applied Mathematics, 12,* 428-454.

Barndorff-Nielsen, O. E., Jensen, J. L. and Kendall, W. S. (Eds.) (1993) *Networks and Chaos - Statistical and Probabilistic Aspects.* London: Chapman and Hall.

Basilevsky, A. and Hum, D. (1977) Spectral Analysis of Demographic Time Series: A Comparison of two Spectral Models. *In:* Barra, J. R., Brodeau, F., Romier, G. and van Cutsem, B. (Eds.) *Recent Developments in Statistics.* Amsterdam: North-Holland.

Berliner, L. M. (1992) Statistics, Probability and Chaos. *Statistical Science, 7,* 69-122.

Berthet, R., Petrossian, A., Residori, S., Roman, B. and Fauve, S. (2003) Effect of multiplicative noise on parametric instabilities. *Physica D, 174,* 84-99.

Bickel, P. J., Ritov, Y. and Rydén, T. (2002) Hidden Markov model likelihoods and their derivatives behave like i.i.d. ones. *Annales de l'Institut Henri Poincaré, PR 38, (6),* 825-846.

Blanchard, F., Maass, A. and Nogueira, A. (Eds.) (2000) *Topics in Symbolic Dynamics and Applications. London Mathematical Society Lecture Note Series, No. 279.* Cambridge: Cambridge University Press.

Blei, R. (2001) *Analysis in integer and fractional dimensions.* [*Cambridge Studies in advanced mathematics, 71.*] Cambridge: Cambridge University Press.

Blei, R. and Janson, S. (2004) Rademacher chaos: tail estimates versus limit theorems. *Archiv für Mathematik, 42,* 13-29.

Bohner, M. and Peterson, A. (2001) *Dynamic Equations on Time Scales.* Boston: Birkhäuser.

Boon, J. P. and Prigogine, I. (1998, personal communication) Le temps dans la forme musicale.

Box, G. E. P. and Jenkins, G. M. (1970) *Time series analysis, forecasting and control.* San Francisco: Holden-Day.

Bowen, R. (1970) Markov partitions for Axiom A diffeomorphisms. *American Journal of Mathematics, 92,* 725-747.

Bower, G. H. and Theios, J. (1964) A learning model for discrete performance levels. *In:* Atkinson, R. C. (Ed.) *Studies in Mathematical Psychology.* Stanford, Cal.: Stanford University Press. 1-31.

Breakspear, M. (2000) Non-linear coupling underlies alpha rhythm dynamics in human EEG. *International Journal of Psychophysiology, 35,* 22.

Brockwell, P. J. and Davis, R. A. (2002) *Introduction to Time Series and Forecasting, 2nd Edn.* New York: Springer-Verlag.

Brucoli, M., Carnimeo, L. and Grassi, G. (1995) Discrete-Time Cellular Neural Networks for Associative Memories with Learning and Forgetting Capacities. *IEEE Transactions on Circuits and Systems - 1: Fundamental Theory and Applications, 42,* 396-399.

Bruhn, J., Ropcke, H., Rehberg, B., Bouillon, T. and Hoeft, A. (2000) EEG approximate entropy correctly classifies the occurrence of burst suppression pattern in anesthesiology. *Anesthesiology, 93,* 981-985.

Buescu, J. (1997) *Exotic Attractors.* Basel: Birkhäuser Verlag.

Bühler, W. K. (1986) *Gauss: Eine biographische Studie.* Berlin: Springer-Verlag.

Buljan, H. and Paar, V. (2002) Parry measure and the topological entropy of chaotic repellers embedded within chaotic attractors. *Physica D, 172,* 111-123.

Bunde, A., Kropp, J, and Schellnhuber, H. J. (2002) *The Science of Disasters.* New York: Springer-Verlag.

Burnham, K. P. and Anderson, D. R. (2002) *Model Selection and Multimodal Inference. (2nd. Edn.* New York: Springer.

Caianiello, E. R. (Ed.) (1987) *Physics of Cognitive Processes.* Singapore: World Scientific.

Caianiello, E. R. (1989) A study of neuronic equations. *In:* Taylor, J. G. and Mannion, C. L. T. *New Developments in Neural Computing.* Bristol: Adam Hilger. 187-199.

Cao, L., Mees, A. and Judd, K. (1998) Dynamics from multivariate time series. *Physica D, 121,* 75-88.

Carvalho, M. P. de (2002) Chaotic Newton's sequences. *The Mathematical Intelligencer, 24,* 31-35.

Chalice, D. R. (2001) How to Differentiate and Integrate Sequences. *The American Mathematical Monthly, 108,* 911-921.

Casdagli, M. (1989) Nonlinear prediction of chaotic time series. *Physica D, 35,* 335-356.

Casdagli. M. C. (1992) Recurrence plots revisited. *Physica D, 108,* 12-44.

Casdagli, M. and Eubank, S. (Eds.) (1991) *Nonlinear Modelling and Forecasting.* Reading, MA: Addison-Wesley.

Chavoya-Aceves, O., Angulo-Brown, F. and Piña, E. (1985) Symbolic Dynamics of the Cubic Map. *Physica D, 14,* 374-386.

Chalice, D. R. (2001) How to Differentiate and Integrate Sequences. *The American Mathematical Monthly, 108,* 911-921.

Chua, G. (Ed.) (1999) *Controlling Chaos and Bifurcations in Engineering Systems.* Boca Raton: CRC Press.

Condon, E. U. (1969) *Scientific Study of Unidentifed Flying Objects.* New York: US Air Force.

Costa, M., Goldberger, A. L. and Peng, C.-K. (2002) Multiscale Entropy Analysis of Complex Physiological Time Series. *Physical Review Letters, 89 (6),* 068102-1, 068102-4

Cox, D. R. and Lewis, P. A. W. (1966) *The Statistical Analysis of Series of Events.* London: Methuen.

Crounse, K. R., Chua, L. O., Thiran, P. and Setti, G. (1996) Characterization and dynamics of pattern formation in cellular neural networks. *International Journal of Bifurcation and Chaos, 6,* 1703-1724.

Crouzet, F. (1982) *The Victorian Economy.* London: Methuen.

Cundy, H. M. (2000) Snubbing with and without eta. *The Mathematical Gazette, 84,* 14-23.

Cutler, C. D. and Kaplan, D. T. (1996) *Nonlinear Dynamics and Time Series: Building a Bridge between the Natural and Statistical Sciences.* Providence, RI: American Mathematical Society.

Cveticanin, L. (1996) Adiabatic Invariants for strongly nonlinear dynamical systems described with complex functions. *Quarterly of Applied Mathematics, LIV (3),* 407-421.

Daubechies, I. (1992) *Ten Lectures on Wavelets.* Philadelphia: SIAM.

Davies, B. (1999) *Exploring Chaos.* Reading, Mass: Perseus Books.

Dawson, S. P. Grebogi, C., Yorke, J. A., Kan, I. and Kocak, H. (1992) Antimonotonicity: Inevitable reversals of period doubling cascades. *Physics Letters A, 162,* 249-254.

de Leeuw, J. (1992) Introduction to Akaike (1973) Information Theory and an Extension of the Maximum Likelihood Principle. *In:* Kotz, S. and Johnson, N. (Eds.) *Breakthroughs in Statistics, Volume I: Foundations and Basic Theory.* New York: Springer-Verlag. 599-609.

Demazure, M. (2000) *Bifurcations and Catastrophes: Geometry of Solutions to Nonlinear Problems.* Berlin: Springer-Verlag.

Dickinson, A. (2001) The 28th Bartlett Memorial Lecture. Causal Learning: An associative analysis. *Quarterly Journal of Experimental Psychology, 54B,* 3-25.

Ding, J. and Fay, T. H. (2005) The Perron-Frobenius Theorem and Limits in Geometry. *The Mathematical Association of America Monthly,* February 2005, 171-175.

Ding, M., Ditto, W. L., Pecora, L. M. and Spano, M. L. (2001) *The 5th Experimental Chaos Conference.* Singapore: World Scientific.

Doyon, B., Cessac, B., Quoy, M. and Samuelides, M. (1993) Control of the transition to chaos in neural networks with random connectivity. *International Journal of Bifurcation and Chaos, 3,* 279-291.

Dunin-Barkowski, W. L., Escobar, A. L., Lovering, A. T. and Orem, J. M. (2003) Respiratory pattern generator model using Ca^{++}-induced Ca^{++} release in neurons shows both pacemaker and reciprocal network properties. *Biological Cybernetics, 89,* 274-288.

Edelman, G. M. and Tononi, G. (2001) *Consciousness: How Matter Becomes Imagination.* London: Penguin Books.

Elliot, R.J., Aggoum, L. and Moore, J. B. (1995) *Hidden Markov Models.* New York: Springer-Verlag.

Engbert, R. and Kleigl, R. (2001) Mathematical models of eye movements in reading: a possible role for autonomous saccades. *Biological Cybernetics, 85,* 77-87.

Engel, A. K., König, P. and Singer, W. (1991) Direct physiological evidence for scene segmentation by temporal coding. *Proceedings of the National Academy of Sciences of the USA, 88,* 9136-9140.

Eykhoff, P. (1974) *System Identification: Parameter and State Estimation.* New York: John Wiley.

Falmagne, J.-C. (2002) *Elements of psychophysical theory.* Oxford: Oxford University Press.

Fechner, G. T. (1860) *Elemente der Psychophysik.* Leipzig: Breitkopf und Härtel.

Feder, J. (1988) *Fractals.* New York: Plenum Press.

Feng, G. (2003) From eye movement to cognition: Toward a general framework of inference. Comment on Liechty et al., 2003. *Psychometrika, 68,* 551-556.

Fischer, P. and Smith, W. R. (Eds.) (1985) *Chaos, Fractals and Dynamics. [Lecture Notes in pure and applied mathematics, vol. 98]* New York: Marcel Dekker.

Fisher, G. H. (1967) Measuring ambiguity. *American Journal of Psychology, 30,* 541-547.

Fokianos, K. and Kedem, B. (2003) Regression Theory for Categorical Time Series. *Statistical Science, 18,* 357-376.

Frank, T. D., Dafferthofer, A., Peper, C. E. Beek, P. J. and Haken, H. (2000) Towards a comprehensive theory of bain activity: Coupled oscillator systems under external forces. *Physica D, 144,* 62-86.

Freeman, W. J. (1994) Characterizations of state transitions in spatially distributed, chaotic, nonlinear dynamical systems in cerebral cortex. *Integrative Physiological and Behavioral Science, 29,* 296-306.

Freeman, W. J. (1999) Noise-induced first-order phase transitions in chaotic brain activity. *International Journal of Bifurcation and Chaos, 9,* 2215-2218.

Frey and Sears (1978) Model of conditioning incorporating the Rescorla-Wagner associative axiom, a dynamic attention process, and a catastrophe rule. *Psychological review, 85,* 321-340.

Friedrich, R. and Uhl. C. (1992) Synergetic analysis of human electroencephalograms: Petit-mal epilepsy. *In:* Friedrich, R. and Wunderlin, A. (Eds.) *Evolution of Dynamical Structures in Complex Systems.* Springer Proceedings in Physics, Vol 69. Berlin: Springer-Verlag.

Friston, K. (2005) A theory of cortical responses. *Philosophical Transactions of the Royal Society, Section B., 360,* 815-836.

Frobenius, G. (1912) Über Matrizen aus nicht negativen Elementen. *Sitzungsberichter der Königlich Preussichen Akademie der Wissenschaften zu Berlin*, 456-477.

Fuentes, M. (2002) Spectral methods for nonstationary spatial processes. *Biometrika, 89*, 197-210.

Fujisaka, H. (1983) Statistical dynamics generated by fluctuations of Local Lyapunov exponents. *Progress in Theoretical Physics, 70*, 1264-1272.

Geake, J. G. and Gregson, R. A. M. (1999) Modelling the internal generation of rhythm as an extension of nonlinear psychophysics. *Musicae Scientiae, 3*, 217-235.

Geritz, S. A. H., Gyllenberg, M., Jacobs, F. J. A. and Parvinen, K. (2002) Invasion dynamics and attractor inheritance. *Journal of Mathematical Biology, 44*, 548-560.

Gillespie, D. T. (1978) Master equations for random walks with arbitrary pausing time distributions. *Physical Letters A, 64*, 22-24.

Gillespie, D. T. (1992) *Markov Processes*. Boston: Academic Press.

Globus, G. (1995) *The Postmodern Brain*. Amsterdam: John Benjamins.

Goldberger, A. L. (1990) Fractal Electrodynamics of the Heartbeat. *In:* Jalife, J. (Ed.) *Mathematical Approaches to Cardiac Arrhythmias. Annals of the New York Academy of Sciences, vol. 591*. New York: New York Academy of Sciences. 402-409.

Gorman, J. M. and Sloan, R. P. (2000) Heart rate variability in depressive and anxiety disorders. *American Heart Journal, 140*, 577-583.

Gopnik, A., Sobel, D. M., Schulz, L.E., and Glymour, C. (2001) Causal learning mechanisms in very young children. Two-, Three-, and Four-year olds infer causal relations from patterns of variation and covariation. *Developmental Psychology, 37*, 620-629.

Gopnik, A., Glymour, C., Sobel, D. M., Schulz, L. E., Kushnir, T. and Danks, D. (2004) A Theory of Causal Learning in Children: Causal Maps and Bayes Nets. *Psychological Review, 111*, 3-32.

Graczyk, J. and Świątek, G. (1998) *The Real Fatou Conjecture. Annals of Mathematics Studies, No. 144*. Princeton, NJ.: Princeton University Press.

Granger, C. W., Maasoumi, E. and Racine, J. (2004) A dependence metric for possibly nonlinear processes. *Journal of Time Series Analysis, 25*, 649-

669.

Grassberger, P., Hegger, R., Kantz, H., Schaffrath, C. and Schreiber, T. (1992) On noise reduction methods for chaotic data. *Chaos, 3,* 127-143.

Green, P. J. (1995) Reversible jump Markov chain Monte Carlo computation and Bayesian model determination. *Biometrika, 82,* 711-732.

Gregson, E. D. and Gregson, R. A. M. (1999) Significant Time Points and Apocalyptic Beliefs. *Australian Folklore, 14,* 170-182.

Gregson, R. A. M. (1983) *Time Series in Psychology.* Hillsdale, NJ: Erlbaum.

Gregson, R. A. M. (1984) Invariance in time series representations of 2-input 2-output psychophysical experiments. *British Journal of Mathematical and Statistical Psychology, 37,* 100-121.

Gregson, R. A. M. (1988) *Nonlinear Psychophysical Dynamics.* Hillsdale, NJ: Erlbaum.

Gregson, R. A. M. (1992) *n-Dimensional Nonlinear Psychophysics.* Hillsdale, NJ: Lawrence Erlbaum Associates.

Gregson, R. A. M. (1993) Learning in the context of nonlinear psychophysics: The Gamma Zak embedding. *British Journal of Mathematical and Statistical Psychology, 46,* 31-48.

Gregson, R. A. M. (1995) *Cascades and Fields in Perceptual Psycho-physics.* Singapore: World Scientific.

Gregson, R. A. M. (1996) n-Dimensional Nonlinear Psychophysics: Intersensory Interaction as a Network at the Edge of Chaos. *In:* MacCormac, E. and Stamenov, M. I. (Eds.) *Fractals of Brain, Fractals of Mind.* Amsterdam: John Benjamins. 155-178.

Gregson, R. A. M. (1997) Intra- and Inter-level Dynamics of Models in Cognitive Neuropsychology. *Proceedings of the 4th Conference of the Australasian Cognitive Science Society,* Symposium on Nonlinear Models in Cognition. Newcastle, NSW: University of Newcastle.

Gregson, R. A. M. (1998a) Herbart and "Psychological Research on the Strength of a Given Impression". *Nonlinear Dynamics, Psychology and Life Sciences, 2,* 157-168.

Gregson, R. A. M. (1998b) Effects of random noise and internal delay on nonlinear psychophysics. *Nonlinear Dynamics, Psychology and Life Sciences, 2,* 73-93.

Gregson, R. A. M. (1999) Analogs of Schwarzian derivatives in NPD, and some nonlinear time series compared. *Unpublished seminar paper, Division of Psychology, The Australian National University.*

Gregson, R. A. M. (2000a) Hierarchical identification of nonlinear psychophysical dynamics in nonstationary realizations. *Cognitive Processing, 1,* 44-51.

Gregson, R. A. M. (2000b) Elementary identification of nonlinear trajectory entropies. *Australian Journal of Psychology, 52,* 94-99.

Gregson, R. A. M. (2000c) Some empirical determinations of a bivariate entropy analogue of the Schwarzian derivative. *National Mathematics Symposium on Nonlinear Time Series, Stochastic Networks and Allied Modern Statistical Techniques.* Canberra: Australian National University Centre for Mathematics and its Applications.

Gregson, R. A. M. (2001a) Responses to Constrained Stimulus Sequences in Nonlinear Psychophysics. *Nonlinear Dynamics, Psychology and Life Sciences, 5,* 205-222.

Gregson, R. A. M. (2001b) Nonlinearity, Nonstationarity, and Concatenation: The Characterisation of Brief Time Series. Invited address to the 10th Australian Mathematical Psychology Conference; Newcastle University, NSW, December 2001.

Gregson, R. A. M. (2002) Scaling Quasiperiodic Psychological Functions. *Behaviormetrika, 29,* 41-57.

Gregson, R. A. M. (2003) Cardiac Slow Dynamic Series. *Paper to the 11th Australian Mathematical Psychology Conference, Sydney, November 2003.*

Gregson, R. A. M. (2004) Transitions between two Pictorial Attractors. *Nonlinear Dynamics, Psychology and Life Sciences, 8,* 41-63.

Gregson, R. A. M. (2005) Identifying ill-behaved Nonlinear Processes without Metrics: Use of Symbolic Dynamics. *Nonlinear Dynamics, Psychology and Life Sciences, 9* .

Gregson, R. A. M. and Harvey, J. P. (1992) Similarities of low-dimension-al chaotic auditory attractor sequences to quasirandom noise. *Perception and Psychophysics, 51,* 267-278.

Gregson, R. A. M. and Leahan, K. (2003) Forcing function effects on nonlinear trajectories: Identifying very local brain dynamics. *Nonlinear Dy-*

namics, *Psychology and Life Sciences, 7,* 137-157.

Gregson, R. A. M. (2005) Identifying ill-behaved Nonlinear Processes without Metrics: Use of Symbolic Dynamics. *Nonlinear Dynamics, Psychology and Life Sciences, 9,* 479-504.

Grove, E. A., Kent, C., Ladas, G. and Radin, M. A. (2001) On $x_{n+1} = max\{\frac{1}{x_n}, \frac{A_n}{x_{n-1}}\}$ with Period 3 Parameter. *Fields Institute Communications, 29,* 161-180.

Guastello, S. J. (1995) *Chaos, catastrophe, and human affairs: Applications of nonlinear dynamics to work, organizations, and social evolution.* Mahwah, NJ: Lawrence Erlbaum Associates.

Guastello, S. J. (2000) Symbolic Dynamic Patterns of Written Exchanges: Hierarchical Structures in an Electronic Problem Solving Group. *Nonlinear Dynamics, Psychology and Life Sciences, 4,* 169-188.

Guastello, S. J., Hyde, T., and Odak, M. (1998) Symbolic dynamic patterns of verbal exchange in a creative problem solving group. *Nonlinear Dynamics, Psychology, and Life Sciences, 2,* 35-58.

Guastello, S. J., Nielson, K. A. and Ross. T. J. (2002) Temporal Dynamics of Brain Activity in Human Memory Processes. *Nonlinear Dynamics, Psychology, and Life Sciences, 6,* 323-334.

Guckenheimer, J. (2003) Bifurcation and degenerate decomposition in multiple time scale dynamical systems. *In:* Hogan, J., Champneys, A., Krauskopf, b., di Bernardo, M., Wilson, E., Osinga, H. and Homer, M. *Nonlinear Dynamics and Chaos: Where do we go from here?* Bristol: Institute of Physics. 1-20.

Gumowski, I. and Mira, C. (1980) *Recurrences and Discrete Dynamic Systems. Lecture Notes in Mathematics, # 809.* Berlin: Springer-Verlag.

Haken, H. and Stadler, M. (Eds.) (1990) *Synergetics of Cognition.* Berlin: Springer-Verlag.

Hale, J. K. (1963) *Oscillations in Nonlinear Systems.* New York: McGraw Hill.

Halgren, E. and Chauvel, P. (1996) Contributions of the human hippocampal formation to the definition of mental contents. *In:* Kato, N. (Ed.) *The Hippocampus: Functions and Clinical Relevance.* Amsterdam: Elsevier. 377-385.

Han, S. K., Kim, W. S., and Kook, H. (2002) Synchronization and Decoding Interspike Intervals. *International Journal of Bifurcation and Chaos*.

Hannan, E. J. (1967) *Time Series Analysis*. London: Methuen.

Hanson and Timberlake (1983) Regulation during challenge: A general model of learned performance under schedule restraint. *Psychological Review, 90,* 261-282.

Hao, Bai-Lin. (1984) *Chaos*. Singapore: World Scientific.

Heath, R. A. (1979) A model for discrimination learning based on the Wagner-Rescorla model. *Australian Journal of Psychology, 31,* 193-199.

Heath, R. A. (1992) A general nonstationary diffusion model for two-choice decision-making. *Mathematical Social Sciences, 23,* 283-309.

Heath, R. A. (2000) *Nonlinear Dynamics: Techniques and Applications in Psychology*. Mahwah, NJ: Erlbaum.

Heath, R. A. (2002) Can people predict chaotic sequences? *Nonlinear Dynamics, Psychology, and Life Sciences, 6,* 37-54.

Heim, R. (1978) On constrained transinformation functions. *In:* Heim, R. and Palm, G. (Eds.) *Theoretical Approaches to Complex Systems. Lecture Notes in Biomathematics, 21*. Berlin: Springer-Verlag.

Helmuth, L. L., Mayr, U. and Daum, I. (2000) Sequence learning in Parkinson's disease; a comparison of spatial-attention and number-response sequences. *Neuropsychologia, 38,* 1443-1451.

Helson, H. (1964) *Adaptation-Level Theory*. New York: Harper and Row.

Hennion, H. and Hervé, L. (2001) *Limit Theorems for Markov Chains and Stochastic Properties of Dynamical Systems by Quasi-Compactness*. [Lecture Noteds in Mathematics, 1766]. Berlin: Springer-Verlag.

Henson, R. (2005) What can functional neuroimaging tell the experimental psychologist? *Quarterly Journal of Experimental Psychology, 58A,* 193-234.

Heyde, C. (2000) Empirical realities for a minimal description risky assets model: the need for fractal features. *Paper at the National Mathematics Symposium on Non-linear Time Series*. Canberra: The Australian National University.

Hilger, S. (1988) *Ein Maßkettenkalkül mit Anwendung auf Zentrumsmannigfaltigkeiten*. Ph.D. thesis, Universität Würzburg.

Hirose, A. (Ed.) (2003) *Complex-Valued Neural Networks.* Singapore: World Scientific.

Honerkamp, J. (1994) *Stochastic Dynamical Systems.* Weinheim: VCH Verlagsgesellschaft mbH.

Hoppensteadt, F. C. (Ed.) (1979) *Nonlinear Oscillations in Biology. (Lectures in Applied Mathematics, vol.17)* Providence, RI: American Mathematical Society.

Hoppensteadt, F. C. (2000) *Analysis and Simulation of Chaotic Systems, 2nd Edn. (Applied Mathematical Sciences, vol. 94.)* New York: Springer-Verlag.

Hoppensteadt, F. C. (2003). Random Perturbations of Volterra Dynamical Systems in Neuroscience. *Scientiae Mathematicae Japonicae, 58,* 353-358.

Hoppensteadt, F. C. and Izhikevich, E. M. (1997) *Weakly Connected Neural Networks. (Applied Mathematical Sciences, vol. 126.)* New York: Springer-Verlag.

Horbacz, K. (2004) Random dynamical systems with jumps. *Journal of Applied Probability, 41,* 890-910.

Horst, P. (1963) *Matrix Algebra for Social Scientists.* New York: Holt, Rinehart and Winston.

Huisinga, W., Meyn, S. and Schütte, C. (2004) Phase transition and metastability in Markovian and molecular systems. *The Annals of Applied Probability, 14,* 419-458.

Infeld, E. and Rowlands, G. (2000) *Nonlinear Waves, Solitons and Chaos. 2nd. Edn.* Cambridge: Cambridge University Press.

Ilyashenko, Yu. and Li, W. (1999) *Nonlocal bifurcations. [Mathematical Surveys and Monographs, 66.]* Providence, RI: American Mathematical Society.

Jarrett, R. G. (1979) A note on the intervals between coal-mining disasters. *Biometrika, 66,* 191-193.

Jensen, O. and Lisman, J. E. (1996) Theta/Gamma networks with Slow NMDA Channels Learn Sequences and Encode Episodic Memory: role of NMDA channels in recall. *Learning and Memory, 3,* 264-278.

Jensen, O., Idiart, M. A. P. and Lisman, J. E. (1996) Physiologically realistic formation of autoassociative memory in networks with theta/gamma oscillations: role of fast NMDA channels. *Learning and Memory, 3,* 243-256.

Jiménez-Montaño, M. A., Feistel, R. and Diez-Martinez,, O. (2004) Information Hidden in Signals and Macromolecules I: Symbolic Time-Series Analysis. *Nonlinear Dynamics, Psychology and Life Sciences, 8,* 445-478.

Johnstone, I. M. and Silverman, B. W. (2004)   Needles and straw in haystacks: Empirical Bayes estimates of possible sparse sequences. *Annals of Statistics, 32,* 1594-1649.

Kaiser, G. (1994) *A Friendly Guide to Wavelets.* Boston: Birkhäuser.

Kalitzin, S., van Dijk, B. W., Spekreije, H. and van Leeuwen, W. A. (1997) Coherency and connectivity in oscillating neural networks: linear partialization analysis. *Biological Cybernetics, 76,* 73-82.

Kaneko, K. and Tsuda, I. (2003) Chaotic Itinerancy. *Chaos, 13,* 926-936.

Kantz, H. and Schreiber, T. (1997) *Nonlinear Time Series Analysis.* Cambridge: Cambridge University Press.

Kapitaniak, T. (1990) *Chaos in Systems with Noise, 2nd Edn.* Singapore: World Scientific.

Kareev, Y. (1992) Not that Bad after All: Generation of Random Sequences. *Journal of Experimental Psychology: Human Perception and Performance, 18,* 1189-1194.

Karpelevich, F. I. (1951) On the characteristic roots of matrices with non-negative elements. [in Russian]. *Izvestiya Akademia Nauk S.S.S.R., seriya mathematicheskogo, 15,* 361-383.

Katok, A. and Hasselblatt, B. (1995) *Introduction to Modern Theory of Dynamical Systems.* New York: Cambridge University Press.

Kauffman, L. and Sabelli, H. (1998) The Process Equation. *Cybernetics and Systems, 29,* 345-362.

Kaufmann, S. A. (2000) *Investigations.* New York: Oxford University Press.

Kawachi, I., Sparrow, D. and Vokonas, P. S. (1994) Symptoms of anxiety and risk of coronary disease. *Circulation, 90,* 2225-2229.

Kay, J. and Phillips, W. A. (1996) Activation functions, computational goals and learning rules for local processors with contextual guidance. *Technical Report No. 96-6.* Department of Statistics, University of Glasgow.

Kaymakçalan, B., Lakshmikantham, V. and Sivasundaram, S. (1996). *Dynamic Systems on Measure Chains.* vol 370 of *Mathematics and its Applications.* Dordrecht: Kluwer Academic.

Kendall, M. G . (1973) *Time-Series.* London: Charles Griffin.

Killeen, P. R. (1989) Behaviour as a Trajectory through a Field of Attractors. *In:* Brink, J. R. and Haden, C. R. (Eds.) *The Computer and the Brain: Perspectives on Human and Artifical Intelligence.* Amsterdam: North Holland. 53-82.

Koch, C. (1997) Computation and the Single Neuron. *Nature, 385,* 207-210.

Krantz, D. H., Luce, R. D., Suppes, P. and Tversky, A. (1971) *Foundations of Measurement, Vol. I.* New York: Academic Press.

Kitoh, S., Kimura, M., Mori, T., Takezawa, K. and Endo, S. (2000) A fundamental bias in calculating dimensions from finite data sets. *Physica D, 141,* 171-182.

Kocic, V. L. and Ladas, G. (1993) *Global Behavior of Nonlinear Difference Equations of Higher Order with Applications.* Amsterdam: Kluwer Academic

Kohmoto, M. (1988) Entropy function for multifractals. *Physical Review, A 37,* 1354-1358.

Kreuzer, E. (1987) *Numerische Untersuchungen nichtlinearer dynamischer Systeme.* Berlin: Springer-Verlag.

Krüger, T. and Troubetskoy, S. (1992) Markov partitions and shadowing for non-uniformly hyperbolic systems with singularities. *Ergodic Theory and Dynamical Systems,12,* 487-508.

Krut'ko, P. D. (1969) *Statistical Dynamics of Sampled Data Systems.* London: Iliffe.

Lansner, A. Associative Processing in Brain Theory and Artificial Intellingence. *In:* Palm, G. and Aertsen, A. *Brain Theory.* Berlin: Springer-Verlag. 193-210.

Large, E. W. and Jones, M. R. (1999) The Dynamics of Attending : How People Track Time-Varying Events. *Psychological Review, 106,* 119-159.

Lathrop, D. P. and Kostelich, E. J. (1989a) Characterization of an experimental strange attractor by periodic orbits. *Physical Review A, 40,* 4028-4031.

Lathrop, D. P. and Kostelich, E. J. (1989b) Analyzing periodic saddles in experimental strange attractors. *In:* Abraham, N. B., Albano, A. M., Passamante, A. and Rapp, P. E. *Measures of Complexity and Chaos.* New York: Plenum Press. 147-154.

Lauwerier, H. A. (1981) *Mathematical Models of Epidemics.* Amsterdam: Mathematisch Centrum.

Lawrence, S., Tsoi, A. C. and Giles, C. Lee (1996) Noisy Time Series Prediction using Symbolic Representation and Recurrent Neural Network Grammatical Inference. Technical Report UMIACS-TR-96-27 and CS-TR-3625, Institute for Advanced Computer Studies. College Park, MD.: University of Maryland.

Liao, H.-E. and Sethares, W. A. (1995) Suboptimal Identification of Nonlinear ARMA Models Using an Orthogonality Approach. *IEEE Transactions on Circuits and Systems - 1: Fundamental Theory and Applications, 42,* 14-22.

Licklider, J. C. R. (1960) Quasi-linear Operator Models in the Study of Manual Tracking. *In:* Luce, R. D., Bush, R. R. and Licklider, J. C. R. (Eds.) *Developments in Mathematical Psychology, Information, Learning and Tracking.* Glencoe, Ill: Free Press. 171-279.

Lind, D. and Marcus, B. (1995) *An Introduction to Symbolic Dynamics and Coding.* Cambridge: Cambridge University Press.

Llinas, R. and Ribary, U. (1993) Coherent 40-hz oscillation characterises dream state in humans. *Proceedings of the National Academy of Science, USA, 90,* 2078-2081.

Luce, R. D. (1959) *Individual Choice Behaviour.* New York: John Wiley.

Luce, R. D., Bush, R. R. and Galanter, E. (1963) *Handbook of Mathematical Psychology, Vol. I.* New York: John Wiley.

Luce, R. D. and Galanter, E. (1963) Discrimination, section 2., Fechnerian Scaling. *In:* Luce, R. D., Bush, R.R. and Galanter, E. *Handbook of Mathematical Psychology, Vol. I* New York: John Wiley. 206-213.

Luo, D. and Libang, T. (1993) *Qualitative Theory of Dynamical Systems. (Advanced Series in Dynamical Systems, Vol. 12.)* Singapore: World Scientific.

Lyubich, M. (1993) Milnor's attractors, persistent recurrence and renormalization. *In:* Goldberg, L. R. and Phillips, A. V. (Eds.) *Topological Methods in Modern Mathematics.* Houston: Publish or Perish, Inc. 513-541.

MacDonald, I. L. and Zucchini, W. (1997) *Hidden Markov and Other Models for Discrete-valued Time Series.* London: Chapman and Hall.

Machens, C. K. (2002) Adaptive sampling by information maximization. *Physical Review Letters, 88,* 228104-1, 228104-4.

Machens, C. K., Romo, R. and Brody, C. D, (2005) Flexible Control of Mutual Inhibition: A Neural Model of Two-Interval Discrimination. *Science, 307,* 1121-1124.

Maguire, B. A., Pearson, E. S. and Wynn, A. H. A. (1952) The time intervals between industrial accidents. *Biometrika, 39,* 168-180.

Maguire, B. A., Pearson, E. S. and Wynn, A. H. A. (1953) Further notes on the analysis of accident data. *Biometrika, 40,* 213-216.

Margraf, J. and Ehlers, A. (1992) Etiological Models of Panic - Medical and Biological Aspects. *In:* Baker, R. (Ed.) *Panic Disorder: Theory Research and Therapy.* New York: John Wiley and Sons. 145-203.

Markus, M., Müller, S.C. and Nicolis, G. (Eds.) (1988) *From Chemical to Biological Organization.* Berlin: Springer-Verlag.

Marmarelis, P. Z. and Marmarelis, V. Z. (1978) *Analysis of Physiological Systems.* New York: Plenum Press.

Marinov, S. A. (2004) Reversed dimensional analysis in psychophysics. *Perception and Psychophysics, 66,* 23-37.

Martelli, M. (1992) *Discrete Dynamical Systems and Chaos. [Pitman Monographs and Surveys in Pure and Applied Mathematics, vol. 62.]* London: Longman Scientific and Technical.

Martin-Löf, P. (1969). The literature on von Mises' kollectives revisited. *Theoria, 35,* 12-37.

May, T. (1987) *An Economic and Social History of Britain, 1760-1970.* Harlow, Essex: Longman.

Mayer-Kress, G. (Ed.) (1986) *Dimensions and Entropies in Chaotic Systems.* Berlin: Springer-Verlag.

McCabe, B. P. M., Martin, G. M. and Tremayne, A. R.(2005) Assessing persistence in discrete nonstationary time-series models. *Journal of Time Series Analysis, 26,* 305-317.

McCord, C. and Mischaikow, K. (1992) Connected simple systems, transition matrices and heteroclinic bifurcations. *Transactions of the American Mathematical Society, 333,* 397-421.

Medvinsky, A. B., Rusakov, A. V., Moskalenko, A. V., Fedorov, M. V. and Panfilov, A. V. (2003) Autowave Mechanisms of Electrocardiographic Variability during High-Frequency Arrhythmias: A Study by Mathematical Modeling. *Biophysics, 48,* 297-305.

Mees, A. (Ed.) (2000) *Nonlinear Dynamics and Statistics.* Boston: Birkhauser.

Meinhardt, G. (2002) Learning to discriminate simple sinusoidal gratings is task specific. *Psychological Research, 66,* 143-156.

Metzger, M. A. (1994) Have subjects been shown to generate chaotic numbers? *Psychological Science, 5,* 111-114.

Meyn, S. P. and Tweedie, R. L. (1993) *Markov Chains and Stochastic Stability.* London: Springer-Verlag.

Michell, J. (2002) Stevens's Theory of Scales of Measurement and its Place in Modern Psychology. *Australian Journal of Psychology, 54,* 99-104.

Miller, R. R., Barnet, R. C. and Grahame, N. J. (1995) *Psychological Bulletin, 117,* 363-386.

Milnor, J. (1992) Remarks on Iterated Cubic Maps. *Experimental Mathematics, 1,* 5-24.

Mitchener, W. G. and Nowak, M. A. (2004) Chaos and Language. *Proceedings of the Royal Society of London, Series B, 271,* 701-704.

Miyano, H. (2001) Identification Model Based on the Maximum Information Entropy Principle. *Journal of Mathematical Psychology, 45,* 27-42.

Morariu, V. V., Coza, A., Chis, M. A., Isvoran, A. and Morariu L.-C. (2001) Scaling in Cognition. *Fractals, 9,* 379-391.

Nadel, L. and O'Keefe, J. (1974) The hippocampus in pieces and patches: An essay on models of explanation in physiological psychology. *In:* Bellairs, R. and Gray, E. G. (Eds.) *Essays on the Nervous System. A Festschrift for J. Z. Young.* Oxford: The Claredon Press. 368-390.

Nadel, L. (1989) Computation in the Brain: The Organization of the Hippocampal Formation in Space and Time. *In:* Brink, J. R. and Haden, C. R. (Eds.) *The Computer and the Brain: Perspectives on Human and Artifical Intelligence.* Amsterdam: North Holland. 83-110.

Nekorkin, V. I., Makarov, V. A., Kazantsev, V. B. and Velarde, M. G. (1997) Spatial disorder and pattern formation in lattices of coupled bistable elements. *Physica D, 100,* 330-342.

Neuringer, A. and Voss, C. (1993) Approximating Chaotic behavior. *Psychological Science, 4,* 113-119.

Nicolis, G. and Prigogine, I. (1989) *Exploring Complexity.* New York: Freeman.

Nicolis, G, Cantù, A. G. and Nicolis, C. (2005) Dynamical Aspects of Interaction Networks. *International Journal of Bifurcation and Chaos, 15,* 3467-3480.

Nishiura, Y. (2002) *Far-from-Equilibrium Dynamics. [Translations of Mathematical Monographs, vol. 299]* Providence, RI: American Mathematical Society.

Novak, A. and Vallacher, R. R. (1998) *Dynamical Social Psychology.* New York: Guilford Press.

Ogorzałek, M. J., Galias, Z., Dąbrowski, A. and Dąbrowski, W. (1996) Wave propagation, pattern formation and memory effects in large arrays of interconnected chaotic circuits. *International Journal of Bifurcation and Chaos, 6,* 1859-1871.

Oprisan, S. A. and Canavier, C. C. (2001) The Influence of Limit Cycle Topology on the Phase Resetting Curve. *Neural Computation, 13,* 1027-1057.

Osgood, C. E. (1953) *Method and Theory in Experimental Psychology.* New York: Oxford University Press.

Ott, E. (1993) *Chaos in dynamical systems.* New York: Cambridge University Press.

Ott, E., Sauer, T. and Yorke, J. A. (1994) *Coping with Chaos.* Now York: John Wiley.

Palm, G. and Aertsen, A. (Eds.) (1986) *Brain Theory.* Berlin: Springer-Verlag.

Parry, W. (1964) Intrinsic Markov Chains. *Transactions of the American Mathematical Society, 112,* 55-66.

Parry, W. (1966) Symbolic dynamics and transformations on the unit interval. *Transactions of the American Mathematical Society, 122,* 368-378.

Parzen, E. (Ed.) (1984) *Time Series Analysis of Irregularly Observed Data.* { *Lecture Notes in Statistics, Vol 25.* } Berlin: Springer-Verlag.

Patterson, D. M. and Ashley, R. A. (2000) *A nonlinear time series workshop: A toolkit for detecting and identifying nonlinear serial dependence.* Boston: Kluwer.

Pearl, J. (1988) *Probabilistic Reasoning in Intelligent Systems.* San Mateo, Cal.: Morgan Kaufmann Publishers.

Peng, C.-K,, Hausdorff, J. M. and Goldberger, A. L. (1999) Fractal mechanisms in neural control: Human heartbeat and gait dynamics in health and disease. *In:* Walleczek, J (Ed.) *Nonlinear Dynamics, Self-organization, and Biomedicine.* Cambridge: Cambridge University Press.

Petersen, K., Quas, A. and Shin, S. (2003) Measures of maximal relative entropy. *Ergodic Theory and Dynamical Systems, 23,* 207-223.

Pilgrim, B. Judd, K. and Mees, A. (2002) Modelling the dynamics of nonlinear time series using canonical variate analysis. *Physica D, 170,* 103-117.

Pincus, S. M. (1991) Approximate entropy as a measure of system complexity. *Proceedings of the National Academy of Sciences of the USA, 88,* 2297-2301.

Pincus, S. M. and Singer, B. H. (1996) Randomness and degrees of regularity. *Proceedings of the National Academy of Sciences of the USA, 93,* 2083-2088.

Pincus, D. (2001) A Framework and Methodology for the Study of Nonlinear, Self-Organizing Family Dynamics. *Nonlinear Dynamics, Psychology, and Life Sciences, 5,* 139-173.

Press, W. H., Flannery, B. P., Teukolsky, S. A. and Vetterling, W. T. (1986) *Numerical Recipes: The art of scientific computing.* New York: Cambridge University Press.

Proakis, J. G., Rader, C. M., Ling, F. and Nikias, C¿ L. (1992) Signal analysis with higher-order spectra. *Advanced Digital Signal Processing.* New York: Macmillan. Chapter 9. 550-589.

Rao, C. V., Wolf, D. M., and Arkin, A. P. (2002) Control, exploitation and tolerance of intracellular noise. *Nature, 420,* 231-237.

Raftery, A. E. and Tavaré, S. (1994) Estimation and modelling repeated patterns in high order Markov chains with the mixture transition distribution model. *Applied Statistics, 43,* 179-199.

Rampil, I. J. (1998) A primer for EEG signal processing in anesthesia. *Anesthesiology, 89,* 980-1002.

Rand, D. A. and Young, L.-S. (Eds.) (1981) *Dynamical Systems and Turbulence, Warwick 1980. Lecture Notes in Mathematics, vol. 898* Berlin: Springer-Verlag.

Redmayne, R. A. S. (1945) *The Problem of the Coal Mines.* London: Eyre and Spottiswoode.

Reichenbach, H. (1949) *The Theory of Probability.* Berkeley: University of California Press.

Rescorla, R. A. (2001) Unequal associative changes when excitors and neutral stimuli are conditioned in compound. *Quarterly Journal of Experimental Psychology, 54B,* 53-68.

Rescorla, R. A. and Wagner, A. R. (1972) A Theory of Pavlovian Conditioning: Variations in the Effectiveness of Reinforcement and Nonreinforcement. *In:* Black, A. H .and Prakasy, W. F. (Eds.) *Classical Conditioning II: Current Research and Theory.* New York: Appleton-Century-Croft.

Rescorla, R. A. and Durlach, P. J. (1981) Within-event learning in Pavlovian conditioning. *In:* Spear, N. E. and Miller, R. R. (Eds.) *Information processing in animals: Memory mechanisms.* Hillsdale, NJ: Lawrence Erlbaum Associates. 81-111.

Revonsou, A. and Kamppinen, M. (1994) *Consciousness in Philosophy and Cognitive Neuroscience.*
Hillsdale, NJ.: L. Erlbaum Associates.

Richman, J.S. and Moorman, J. R. (2000) Physiological time series analysis using approximate entropy and sample entropy. *American Journal of Physiology: Heart Circulatory Physiology, 278,* H2039-H2049.

Robert, C. P. and Casella, G. (1999) *Monte Carlo Statistical Methods.* New York: Springer.

Rosenstein, M.T., Collins, J. J. and De Luca, C. J. (1993) A practical method for calculating largest Lyapunov exponents from small data sets. *Physica D, 65,* 117-134.

Rössler, O. E. (1991) The Chaotic Hierarchy. *In:* Baier, G. and Klein, M. (Eds.) *A Chaotic Hierarchy.* Singapore: World Scientific. 25-47.

Roweis, S. T., and Saul, L. T. (2000) Nonlinear Dimensionality Reduction by Locally Linear Embedding. *Science, 290,* 2323-2326.

Ruhnau, E. (1995) Time-Gestalt and the Observer. *In:* Metzinger, T. (Ed.) *Conscious Experience.* Thorverton, UK.: Imprint Academic Schöningh. 165-184.

Salmon, P. (2003) How do we recognise good research? *The Psychologist, 16, (1),* 24-27.

Sandefur, J. T. (1990) *Discrete Dynamical Systems.* Oxford: Clarendon Press.

Schaffer, W. M., Ellner, S. and Koty, M. (1986) Effects of noise on some dynamical models in ecology. *Journal of Mathematical Biology, 24,* 479-523.

Schiff, S. J., So, P., Chang, T., Burke, R. E. and Sauer, T. (1996) Detecting dynamical interdependence and generalized synchrony through mutual prediction in a neural ensemble. *Physical Review E,* 6708-6724.

Schuster, H. G.(1984) *Deterministic Chaos.* Weinheim: Physik-Verlag GmbH.

Schwarz, H. A. (1868)i Über einige Abbildungs aufgaben. *Journal für reine und angewandte Mathematik, 70, (Crelle),* 105-120.

Seneta, E. (1973) *Non-Negative Matrices.* London: George Allen and Unwin.

Setiono R. and Liu, H. (1996) Symbolic Represesentation of Neural Networks. *Computer, 29(3),* 71-77.

Shafer, G. (1994) The Subjective Aspect of Probability. Chapter 4 *In:* Wright, G, and Ayton, P. (Eds.) *Subjective Probability.* Chichester: John Wiley and Sons. 53-74.

Shanks, D. R. (1985) Forward and Backward Blocking in Human Contingency Judgment. *The Quarterly Journal of Experimental Psychology, 37B,* 1-21.

Shilnikov, L. P. (1966) Birth of a periodic movement departing from the trajectory that goes to an equilibrium state of a saddle-point-saddle-point and then returns. [In Russian] *Proceedings of the Academy of Sciences of the USSR, 170,* 49-52.

Shilnikov, L. P., Shilnikov, A. L., Turaev, D. V. and Chua, L. O. (1998) *Methods of Qualitative Theory in Nonlinear Dynamics, Part 1.* Singapore: World Scientific. Chapter 2.

Shye, S. (Ed.) (1978) *Theory construction and data analysis in the behavioral sciences.* San Francisco: Jossey-Bass.

Sil'nikov, L. P. (1965) A case of the existence of a countable number of periodic motions. *Soviet Mathematics, 6,* 163-166. [Translation from *Doklady Akademii Nauk SSSR*].

Sinai, Ya. G.(Ed.) (2000) *Dynamical Systems, Ergodic Theory and Applications.* [*Encyclopaedia of Mathematical Sciences, Volume 100.*] Berlin: Springer-Verlag.

Singer, W. (1993) Synchronization of cortical activity and its putative role in information processing and learning. *Annual Review of Physiology, 55,* 349-3374.

Singer, W. and Gray, C. M. (1995) Visual feature integration in the temporal correlation hypothesis. *Annual Reviews of Neuroscience, 18,* 555-586.

Smithson, M. (1997) Judgment under chaos. *Organizational Behavior and Human Decision Processes, 69,* 59-66.

Sobottka, M. and de Olveira, L. P. L. (2006) *Monthly of the Mathematical Association of America, May 2006,* 415-424.

Spall, J. C. (Ed.) (1988) *Bayesian Analysis of Time Series and Dynamic Models.* New York: Marcel Dekker.

Srinivasan, R., Russell, D. P., Edelman, G. M. and Tonini, G. (1999) Increased Synchronization of Neuromagnetic Responses during Conscious Perception. *Journal of Neuroscience, 19,* 5435-5448.

Stevens, S. S. (Ed.) (1951) *Handbook of Experimental Psychology.* New York: John Wiley.

Strogatz, S. H. (1986) *The Mathematical Structure of the Human Sleep-Wake Cycle.* [*Lecture Notes in Biomathematics, vol. 69]* Berlin: Springer-Verlag.

Sugihara, G. and May, R. M. (1990) Nonlinear forecasting as a way of distinguishing chaos from measurement error in time series. *Nature, 344,* 734-741.

Sullivan, D. Conformal dynamical systems. *In:* Palis, J. (Ed.) *Geometric Dynamics. [Lecture Notes in Mathematics, 1007.]* Berlin: Springer-Verlag. 725-752.

Tass, P., Rosenblum, M. G., Weule, J., Kurths, J., Pikovsky, A., Volkmann, J., Schnitzler, A. and Freund, H.-J. (1998) Detection of $n : m$ Phase Locking from Noisy Data: Applicatin to Magnetoencephalography. *Physical Review Letters, 81*, 3291-3294.

Theiler, J., Eubank, S., Longtin, A., Galdrikian, B. and Farmer, J. D.(1992) Testing for nonlinearity in time series: The method of surrogate data. *Physica D, 58*, 77-91.

Theiler, J., Galdrikian, B., Longtin, A., Eubank, S., and Farmer, J. (1992) Using surrogate data to detect nonlinearity in time series. *In:* Casdagli, M. and Eubank, S. (Eds.) *Nonlinear modelling and forecasting.* Redwood City: Addison-Wesley.

Thelen, E. and Smith, L. B. (1994) *A Dynamical Systems Approach to the Development of Cognition and Action.* Cambridge, Mass.: MIT Press.

Thiran, P., Crounse, K. R., Chua, L. O. and Hasler, M. (1995) Pattern Formation Properties of Autonomous Cellular Neural Networks. *IEEE Transactions on Circuits and Systems - 1: Fundamental Theory and Applications, 42*, 757-774.

Thom. R. (1975) *Structural Stability and Morphogenesis: An outline of a General Theory of Models.* Reading, Mass: Benjamin Cummings.

Thompson, J. R. (1999) *Simulation. A Modeler's Approach.* New York: John Wiley.

Thomsen, J. S., Mosekilde, E. and Sterman, J. D. (1991) Hyperchaotic Phenomena in Dynamic Decision Making. *In:* Mosekilde, E. and Mosekilde, L. (Eds) *Complexity, Chaos and Biological Evolution.* New York: Plenum Press. 397-420.

Tirsch, W. S., Keidel, M., Perz, S., Scherb, H. and Sommer, G. (2000) Inverse covariation of spectral density and correlation dimension in cyclic EEG dynamics of the human brain. *Biological Cybernetics, 82*, 1-14.

Tong, H. (1983) *Threshold Models in Non-linear Time Series Analysis. Lecture Notes in Statistics, 21.* Berlin: Springer-Verlag.

Torres, M. E. and Gamero, L. G. (2000) Relative complexity changes in time series using information measures. *Physica A, 286,* 457-473.

Treffner, P. J. and Turvey, M. T. (1996) Symmetry, broken symmetry, and handedness in bimanual coordination dynamics. *Experimental Brain Research, 107,* 463-478.

Tucker, P., Adamson, P.and Mirander, R. (1997) Paroxetine increases heart rate variability in panic disorder. *Journal of Clinical Psychopharmacology, 17,* 370-376.

Tukey, J. W. (1977) *Exploratory Data Analysis.* Reading, Mass.: Addison-Wesley.

Udwadia, F. E. and Guttalu, R. S. (1989) Chaotic dynamics of a piecewise cubic map. *Physical Review A, 40(7),* 4032-4044.

van der Wollenberg, A. L. (1978) Nonmetric Representation of the Radex in Its Factor Pattern Parametrization. *In:* Shye, S. (Ed.) *Theory construction and data analysis in the behavioral sciences.* San Francisco: Jossey-Bass. 326-349.

van Leeuwen, C. and Raffone, A. (2001) Coupled nonlinear maps as models of perceptual pattern and memory trace dynamics. *Cognitive Processing, 2,* 67-116.

van Strien, S. (1988) Smooth Dynamics on the Interval. *In* Bedford, T. and Swift, J. (Eds.) *New Directions in Dynamical Systems. London Mathematical Society, Lecture Note Series, 127.* 57-119.

Varga, R.S. (1962) *Matrix Iterative Analysis.* New Jersey: Prentice Hall.

Viana, M. (2000) What's new on Lorenz strange attractors?. *The Mathematical Intelligencer, 22,* 6-19.

Vickers, D. (1979) *Decision processes in visual perception.* New York: Academic Press.

Vinje, W. E. and Gallant, J. L. (2000) Sparse coding and decorrelation in primary visual cortex during natural vision. *Science, 287, (18),* 1273-1276.

Visser, I., Raijmakers, M. E. J. and Molenaar, P. C. M. (2000) Confidence intervals for hidden Markov model parameters. *British Journal of Mathematical and Statistical Psychology, 53,* 317-327.

Volchan, S. B. (2002) What is a Random Sequence? *American Mathematical Monthly, 109,* 46-63.

Wagenaar, W. A. (1972) Generation of random sequences by human subjects: A critical survey of literature. *Psychological Bulletin, 77,* 65-72.

Wang, X. and Yang, X. (1992) *Birth and Death Processes and Markov Chains.* Berlin: Springer-Verlag.

Wang, S. Y., Zhang, L. F., Wang, X. B. and Cheng, J. H. (2000) Age dependency and correlation of heart rate variability, blood pressure variability and baroreflex sensitivity. *Journal of Gravitational Physiology, 7,* 145-146.

Ward, L. M. and West, R. L. (1998) Modeling Human Chaotic Behaviour: Nonlinear Forecasting Analysis of Logistic Iteration. *Nonlinear Dynamics, Psychology, and Life Sciences, 2,* 261-282.

Warner, R. M. (1998) *Spectral Analysis of Time-Series Data.* New York: The Guilford Press.

Warren, R. M., Bashford, J. A., Cooley, J. M. and Brubaker, B. S. (2001) Detection of acoustic repetition for very long stochastic patterns. *Perception and Psychophysics, 63,* 175-182.

Weatherburn, C. E. (1927) *Differential Geometry of Three Dimensions.* Cambridge: Cambridge University Press.

West, B. J., Hamilton, P. and West, D. J. (2000) Random Fractal Time Series and the Teen-Birth Phenomenon. *Nonlinear Dynamics, Psychology, and Life Sciences, 4,* 87-112.

White, P. A. (2005) Cue interaction effects in causal judgement: An interpretation in terms of the evidential evaluation model. *The Quarterly Journal of Experimental Psychology, 58B,* 99-140.

Whitfield, J. W. (1950) The Imaginary Questionnaire. *Quarterly Journal of Experimental Psychology, 2,* 76-87.

Wiggins, S. (1988) *Global Bifurcations and Chaos.* New York: Springer-Verlag.

Wiggins, S. (1992) *Chaotic Transport in Dynamical Systems.* New York: Springer-Verlag.

Winfree, A. T. (1987) *When time breaks down.* Princeton: Princeton University Press.

Winkler, G. (2003) *Image Analysis, Random Fileds and Markov Chain Monte Carlo Methods. 2nd. Edn.* Berlin: Springer.

Wolpert, D. H. and Macready, W. G. (1997) No free lunch theorems for search. *IEEE Transactions on Evolutionary Computation, 1,* 67-82.

Wright, J. J. and Liley, D. J. T. (1996) Dynamics of the brain at global and microscopic levels: Neural Networks and the EEG. *Behavioral and Brain Sciences, 19,* 285-320.

Yao, W., Yu, P., Essex, C. and Davison, W. (2006) *International Journal of Bifurcation and Chaos, 16,* 497-522.

Zak, M. (2004) Dynamical networks for information processing. *Information Sciences, 165,* 149-169.

Zaliapin, I., Gabrielov, A. and Keilis-Borok, V. (2004) Multiscale Trend Analysis. *Fractals, 12,* 275-292.

Zbilut, J. P. (2004) *Unstable Singularities and Randomness.* Amsterdam: Elsevier.

Zheng, G., Freidlin, B. and Gastwirth, J. L. (2004) Using Kullback-Leibler information for model selection when the data-generating model is unknown: Applications to genetic testing problems. *Statistica Sinica, 14,* 1021-1036.

Ziehmann, C., Smith, L. A. and Kurths, J. (2000) Localized Lyapunov exponents and the prediction of predictability. *Physics Letters A, 271,* 237-251.

Subject Index